Successful Small Farms
Building Plans & Methods

Successful
Small Farms
Building Plans & Methods

Herbert T. Leavy

Structures Publishing Co. 1978
Farmington, Michigan 48024

Copyright © 1978 Herbert T. Leavy

All rights reserved, including those of translation. This book or parts thereof may not be reproduced in any form without permission of the copyright owner. Neither the author, nor the publisher, by publication of data in this book, ensures to anyone the use of such data against liability of any kind, including infringement of any patent. Publication of data in this book does not constitute a recommendation of any patent or proprietary right that may be involved.

Manufactured in the United States of America

Book edited by Shirley M. Horowitz

Book design by Carey Jean Ferchland

Cover photo courtesy of
Bernie Donahue, Desoto, Missouri—Candida Photos, Inc.

Current printing (last digit)
10 9 8 7 6 5 4 3 2 1

Structures Publishing Co.
Box 1002, Farmington, Mich. 48024

Library of Congress Cataloging in Publication Data

Leavy, Herbert T
   Successful small farms.

   Bibliography: p.
   Includes index.
   1. Farm buildings—Design and construction.
2. Livestock housing.   I. Title.
TH4911.L4   690'.8'9   78-7987
ISBN 0-912336-67-6
ISBN 0-912336-68-4 pbk.

Contents

Introduction 7
SECTION I: **Planning**
    1. Checklists 8
    2. Farmstead Planning 10
    3. Housing for Animals 15

SECTION II: **Setting Up the Basic Farm Structure**
    4. Money Matters 30
    5. Building Basics 32
    6. Environmental Controls 51
    7. All About Roofing Farm Buildings 55
    8. Working with Corrugated Steel 70
    9. Working with Plywood 82
    10. Working with Adobe 98
    11. Interior Wiring 102

SECTION III: **Specialized Needs**
    12. Building Greenhouses (including Plastic-Covered)
        Hotbed and Propagating Frame 113
    13. Movable Shed 132
    14. Housing for Horses 133
        Horse Barn (including Indoor Exercise Area)
        Portable Stable
        Construction Tips
    15. Cattle 149
        Adjustable Cattle Chute
        Wagon Rack For Silage Feeding
    16. Swine 153
        Hog Feeder
        House for Dry Sows
        Farrowing House
    17. Poultry 158

SECTION IV: **Maintenance**
    18. Fences 161
    19. Caulks, Sealants, Putties & Glazing Compounds 174
    20. Building Maintenance & Exterior Paint 178

Appendix A: Simple Methods of Wood Preservative Treatment 183
Appendix A.1: Foundation Footing Depths 184
Appendix B: List of Manufacturers and Their Addresses 185
Index 187

ACKNOWLEDGEMENTS

We wish to thank the following persons and organizations for their enthusiastic help and support in producing this book. Without their cooperation, the book would not have been possible: Mr. Robert J. Schliebe, Granite City Steel; Mr. Karl Dahlem and Mr. Don Bridwell of Wickes Lumber and The Wickes Corporation; Mr. Bern Fleming of the Jim Walter Corporation; Mr. J.S. Titus of W.R. Grace & Co.; Mr. Jon Avrigean of Wheeling Pittsburgh Steel Corporation; Mr. Art Somers of Evans Products Company; Mr. Herb Kinkead of Butler Manufacturing Company; Mr. Jack M. Ullrich of Andersen Corporation; Mr. Thomas A. Newton of CertainTeed Corporation; Mr. Christopher D. Cook of Cook Communications Services for Pascoe Steel Corporation; Mrs. Maryann Ezell of the American Plywood Association; Mr. Augustus Suglia; Mr. John M. Deakins of Fabral Corporation; Ms. Patricia Fowler of the Star Manufacturing Company.

Line art courtesy of: Granite City Steel—pages 33, 36, 37, 38, 39, 41, 42, 43, 44, 46, 52, 53, and 54; Wheeling-Pittsburgh Steel Co.—pages 11, 20, 21, 22, 23, 24, 25, and 26.

DEDICATION

This book is dedicated to my children—
Karen, Kathy, Jacqueline and Jill.

Introduction

Farming for supplemental financial support and food is one of the fastest growing lifestyles today—and farming technology is one of the fastest-changing industries. The modern farmer is substituting machinery and modern know-how for yesterday's back-breaking labor methods.

For the smaller farmer, however, many of the recent innovations requiring large amounts of capital investment are not always the best choice. The planning principles and building programs covered in this book, as well as the technical construction details, will help the reader make his best choice depending upon the type and scale of farming intended.

Several groups of people should benefit from the material presented in this book: (1) those who live on small farms but who also have full-time jobs and thus have a limited operation and limited time; (2) young back-to-the-land aspirants who have a great deal of time and energy but not much money to put into farming facilities; (3) retirees who want to be self-supporting on a small scale; (4) the wealthy, who use the farm as a leisure and vacation area (particularly ranches and horse farms); and (5) those who are engaged in full-time farming on a small farm and want to know how to make improvements and expansions without risking their means of support.

Successful Small Farms—Building Plans & Methods divides the subject matter into four areas: Planning, Building, Specific Projects, and Maintenance. In each section the basics are covered in relation to various types of farming, whether storage or horses or poultry or beef or swine. Each subject, such as temperature and humidity control or feeder operations, is related to the animals involved.

SECTION 1: Planning

1. Checklists

Checklists

Even before you begin planning your building or buildings, you will want to review these checklists, and perhaps add additional categories that will suit your needs.

Questions you may have about the checklists provided here will be answered within the chapters relating to the specific details. The checklists point up what you need to know prior to construction and ordering.

Not all items on these checklists will apply to your building needs, but many will and it will pay you in dollars and time saved if you go through them carefully prior to ordering your plan, or beginning construction.

Design and Engineering Services
 Decide whether or not to employ an architect or engineer.
 Check into their fees.
 Look into standard plans available (see those illustrating this book).
 Check surveys that will be needed: survey of lot, traffic surveys, etc.
 Are soil tests necessary?
 Set up reproduction expenses—plans, printing, copies of land titles, etc.
 Check need for a performance bond to guarantee completion of work in case construction company goes out of business.
 What other types of bonds will be necessary?
 Research your potential insurance needs: builder's risk, contents, workmen's accident, fire, etc.
 Budget your brokerage and legal fees.
 Budget your miscellaneous expenses, such as travel, postage, phone costs.
 Compute a cost estimate for the above.

Off-Site Improvements
 Will the site require a new entrance from the street?
 Will traffic controls be necessary to reach the site?

On-Site Development
 Check the need for demolition work.
 Arrange for clearing, grading, fill or dirt removal.
 Plan parking and traffic facilities, parking area and driveways.
 How many cars should the drive or garage hold?
 How many square feet of parking area will be needed?
 Contact paving contractor.
 Consider delivery parking and driveways.
 Need pedestrian walkways?
 Will retaining walls be required?
 Will new fencing or screening be required?
 What other on-site development will be required?
 Compute an estimated cost for the above.

Exterior Utilities and Connections Up to the Building
 Make arrangements for temporary power to the job site during construction.
 What are the power requirements to the building?
 Need new gas or water mains?
 Make arrangements for storm water and drainage facilities and/or piping of underground streams.
 Arrange for piping from the building to the sewer trunk, septic system or sewer injector, if needed.
 Check on having to relocate existing power and/or phone lines.
 Compute an estimated cost for the above.

Exterior Identification and Lighting
 Check the need for flood lighting for outside areas.
 What parking lot lighting, including installation, fixtures, stanchions, lamps, bumper guards and railings will be required?
 Compute an estimated cost.

Building Shell: General
 What type of excavation and foundation will be required?
 What type of exterior walls will serve best?
 What type of floor will be required?
 What type of roof will be required?
 Check insurance costs for metal roofs versus ordinary.
 What type of insulation will serve best in the walls and roof?
 Consider the architectural look of the building as opposed to your home.
 List special hardware, such as electric door operating equipment, that may be required.
 Research the type of ventilation system required or needed.

Check annual rainfall to choose size of gutters and downspouts.
Select exterior paint—check long term maintenance needs.
Choose colors to provide a pleasing, compatible exterior appearance.
Compute cost.

Building Shell: Flooring, Ceilings, Partitions, Interior Painting
What other floor surfaces will be needed?
What type of ceiling finish will be needed?
Choose the proper type of interior wall partitioning to reduce annual insurance premiums.
What type of interior painting will be best?
Compute cost.

Mechanical and Electrical
Plan interior wiring requirements.
Check on need for special wiring for power tools, etc.
Carefully plan the location of electrical outlets.
Will electrical outlets be mounted in the floor and/or walls?
Provide for sufficient capacity on meters, subpanels and transformers.
Decide general light requirements.
List overall mechanical requirements, if any.
List heating, air conditioning and ventilation requirements.

Plan any waste disposal requirements, such as a trash room or dispose-all.
Will you need a roof sprinkler? Consider a sprinkler system.
Consider the need for a fire or burglar alarm system.
Compute cost for the above.

Landscaping
Develop a plan for landscaping; choose the type and position of trees, shrubbery, flowers.
Will automatic sprinkling system be required?
Use an outside consultant, have the contractor provide this service or do it yourself?
Compute cost of the above.

Contractors
Choose a contractor after comparing estimates and references.
Decide if you want full service, including building, planning, assistance in site selecting, even to landscaping. In short, is a contractor needed who can coordinate and construct the entire building package?
Or, are you planning to coordinate all phases of the construction project and purchase each portion of your building through self-appointed subcontractors?
Check on the quality of the products he specifies to you.
Personally visit some of the jobs he has constructed.

2. Farmstead Planning

The site arrangement of farmstead buildings is very important. Proper building and lot locations increase efficiency and can save time, space, and money.

Few farms actually complete a new farmstead system in one year. Good farmstead planning, therefore, must include consideration for future expansion possibilities.

Plans That Save Space and Time

Good farm management necessitates an efficient layout of farm buildings that conserves space and time. An entire farmstead can seldom be laid out at one time, as most farms already have buildings which will influence the location of additional buildings. However, a total farmstead plan should be established and any new buildings, fences, lots, trees or orchards should be fitted into this plan.

Farmstead layout involves two problems: (1) type of operation and location of the building site on the farm; (2) location and arrangement of individual buildings within the site.

Farm buildings should be located to minimize walking distances and to accomplish the greatest amount of work with the least amount of labor. The average livestock farmer spends 75 percent of his working time around his buildings.

In general, the barn, machine shed, granary, and other service buildings should be located around a central court and so arranged that most of them can be seen from the house. The court should be large enough to provide space to turn and park machinery. The larger the court, however, the more steps are required to do chores, so distance should be kept within reasonable limits.

Farmstead arrangement involves individual problems of wind direction, slope, drainage, and the relation of buildings to fields and roads. Thus, the best plan may be somewhat different on each farm.

The following factors should be used as guides in determining the desirability of a site for the farmstead:

- Topography—High, level, without abrupt slopes
- Drainage—High, nearly level, good outlets, dry lots
- Water Supply—Good spring and easy to obtain wells
- Soil—Porous, dries quickly for lots; loose, friable, productive for garden
- Fields—All close and easy to reach
- View—Long-range scenic view
- Sun and Wind Exposure—Buildings face road to the south or east; prevailing winds carry odors away from house
- Electricity—On good power line; three-phase service available
- Schools and Churches—On bus line or close to school
- Telephone—On well-maintained line

In most cases, the farmstead selected will not meet all of the above requirements. It is best to find a location that will meet most of them. The more good points possessed by a farmstead, the happier and more satisfied the farm family will be.

When the farmstead area has been selected, the work of locating buildings begins. Small scale models or cardboard boxes cut to reflect the sizes of these buildings can be very helpful.

Draw the farmstead plot to the same scale as the building models. Arrange and rearrange the models, to provide visual assistance in arriving at a practical and efficient layout.

Buildings should usually be set so walls of nearby or facing buildings are square and are either parallel or at right angles to each other. This produces an easily comprehensible pattern and allows neat and orderly arrangement of lot fences and gates.

If possible, natural features of topography should be used to shelter the farmstead from heavy winds and drifting snow. If such protection is not available, suitable trees and shrubs should be planted as windbreaks.

Guidelines for Building Location within Farmstead

House should:

(1) command view of other buildings;
(2) be easily accessible;
(3) be set so visitors can go to either front or back door, but first inclination should be to front door;
(4) be located to make maximum use of sunlight;
(5) be at least 100 ft. from road except in heavy snow areas, where it may be less; and, when on a heavily traveled road, 150 ft. is preferable;
(6) have yard adaptable to attractive landscaping;
(7) be located on southwest corner of farmstead.

Layouts that Place Buildings in Proper Relation to House, Prevailing Winds and Access Road

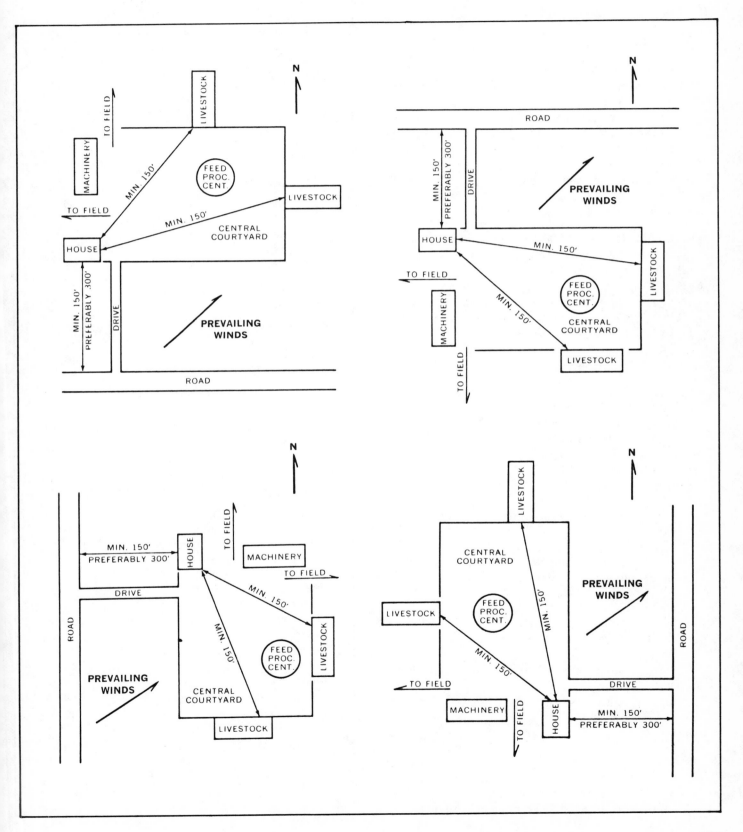

SECTION I: Planning

Garage should:

(1) be close to or attached to house;
(2) be located so snow will not drift in front;
(3) be reasonably close to shop.

Beef and Dairy Facilities should:

(1) have main entrance open to court area;
(2) be located so that the prevailing winds carry odors away from house;
(3) have livestock lot readily accessible to pasture;
(4) have handling pens and loading chute accessible to court;
(5) be accessible to good water supply;
(6) be close to cultivated fields.

Hog Facilities should:

(1) be not closer than 200 feet to home and located so that prevailing winds carry odors away from house;
(2) have one end on court;
(3) be close to grain storage and feed-handling center;
(4) be accessible to a good water supply.
(5) be located far away from any current or previous poultry facilities (to prevent disease).

Sheep Facilities should:

(1) be within clear view of house and preferably not over 200 feet away—and located so that prevailing winds carry odors away from house;
(2) be adjacent to pasture;
(3) be accessible to good water supply.

Poultry Facilities should:

(1) be located so prevailing winds carry odors away from house;
(2) have driveway access at both ends;
(3) be accessible to water and electricity supply;
(4) face south if open-front ventilation is planned.

Grain Storage should:

(1) be centrally located between grain-consuming enterprises;
(2) have adequate open space for filling and drawing of grain;
(3) be readily accessible from the fields.

Silo should:

(1) be close for convenient feeding in either barn or lots;
(2) be arranged so silage goes directly into feed bunk, wagon or silage cart for feeding;
(3) have space around it for equipment at filling time.

Weights of Commodities and Materials

Grains, Feed, and Seeds

Commodity	Pounds Per Bushel	Pounds Per Cubic Feet
Alfalfa meal	19	15.2
Alfalfa seed	60	48
Barley	48	38.4
Blue Grass seed	14	11.2
Brome Grass seed	14	
Clover seed	60	48
Corn (husked ear)	70	28
Corn (shelled)	56	44.8
Corn meal	50	40
Corn & cob meal	45	36
Cotton seed meal	48	38.4
Cotton seed	33	26.4
Oats (whole)	32	25.6
Oats (ground)	22	17.6
Rye	56	44.8
Soybeans	60	48
Soybean oil meal		
Timothy seed	45	36
Wheat	60	48
Wheat bran	16	12.8

Fruits and Vegetables

Commodity	Common Measure	Pounds Per Unit	Pounds Per Cubic Foot
Apples	Northwest box (10½x11½x18)	44	38
	Eastern box (11x13x17)	54	
	Bushel	48	
Beets	Bushel	60	48
Carrots	Bushel	50	40
Molasses	Barrel	650	
Onions	Bushel	57	45.6
Peaches	Bushel	48	36.4
Pears	Bushel	50	40
Peas	Bushel	60	48
Potatoes (Irish)	Bushel	60	48
Potatoes (Sweet)	Bushel	50	40
Tomatoes	Bushel	60	48
Turnips	Bushel	55	44
Cherries	Box (3¾x11½x14⅛)	15	
	Bushel	64	51

Storage Capacities for Trench or Bunker Silos

For computing tons of storage, use the following formulas:

Capacity in cubic feet =
$$\frac{\text{Top width} + \text{bottom width}}{2} \times \text{height} \times \text{length}$$

Capacity in tons, corn or sorghum

(35 pound cubic feet) = $\dfrac{\text{Capacity cubic feet}}{60}$

Capacity in tons grass silage

(40 pound cubic feet) = $\dfrac{\text{Capacity cubic feet}}{50}$

Hay Storage Requirements
(cubic feet per ton)

Baled hay	250
Average chopped hay	250
Fine chopped hay	225
Loose hay	500

Amount of Liquid Manure Produced per Day

Dairy	11 gallons or 1½ cubic feet
Beef	5½ gallons or ¾ cubic foot
Hog	2 gallons or ¼ cubic foot
Poultry	.033 gallon

Machine Storage and Shop should:

(1) be at one side of the central court;
(2) be located so equipment can be driven through and around storage area;
(3) have open storage protected by windbreak or overhang;
(4) be accessible to fields;
(5) be in line of travel between house and other buildings.

Windbreaks add to farm efficiency because they:

(1) reduce extremes of temperatures;
(2) allow greater use of open sheds;
(3) reduce the need for insulation or make it more effective;
(4) reduce damage by strong winds, especially to doors and windows;
(5) reduce exposure and thus add to the service life of buildings and equipment;
(6) improve workability around buildings, which increases the effectiveness of labor.
(7) For maximum effectiveness, extend windbreaks across north and west sides of farmstead and, in far northern areas, around northeast and southwest corners.

Machinery and Equipment Storage

Reliable operation and serviceability of machines are important factors in performing all operations of production. Machines must be ready when a job needs to be done. Farm machinery is more complicated, with more moving and intricate parts, than it was several years ago. Because of this, keeping the machine dry and out of the weather when not in use is increasingly important to maintenance and use. Machine storage buildings are necessary to maintain condition and value.

A new and popular type of storage building is an enclosed structure with doors along one side and at one end for easy drive-through accessibility. Open-front buildings with protective overhang also prove practical.

Does Machinery Storage Pay?

Machinery storage is responsible for savings in repair cost. A survey by Pennsylvania State University revealed that repair costs were reduced 19 percent annually on housed tractors. Repair costs on self-propelled combines were reduced 6 percent annually when stored inside.

Advantages other than repair often justify the cost of a machinery storage building. Machinery with high seasonal use that has been sitting unprotected outside will invariably give more breakdown trouble and will take longer to get ready for operation than if it had been protected by shelter.

Guidelines for Machinery Housing

A practical machine storage building is one in which most of the equipment can be reached without having to move other pieces of equipment. It should be large enough to allow machinery to be stored back-to-back in some areas.

The eave height or end door height should be adequate to accommodate the tallest machine. The front openings or doors should be wide enough for your widest equipment.

Machinery left outside will not return dividends on a vacant building—the building must be easy to use! The only way to make a shelter work full time is to make it just as easy to leave the equipment under the roof as to park it outside.

SECTION I: Planning

Equipping a Farm Shop

The farm shop is an increasingly important facility on the farm. With a properly equipped shop, major repairs can be made as well as necessary minor ones.

The shop should be large enough to hold the largest tools and equipment and still have ample room to move about.

Proper ventilation and lighting are a must in any farm shop, and it is advantageous to have the room heated for winter comfort.

Records of machines are also important so you can determine costs of repairs and maintenance.

Electric Uses and Needs

Estimated Kilowatt-Hour Use by Farm Electric Chore Equipment

General	Average KWH Use	Swine Equipment	Average KWH Use
Conveyor—Auger	4 per 1000 bushel	Brooding (Infrared)	40 per litter
Elevator—Auger	10 per 1000 bushel	Farrowing—Floor Cable	30 per sow farrowed
Bucket	3 per 1000 bushel	Ventilator	1 per month per pig
Feed Grinder (general)	5 per ton	Water fountains (auto)	.5 per month per pig
Mixer (general)	1 per ton	Bunk feeder	1 per month per pig
Feeder (automatic)	2 per ton	Feed grinding	.3 per month per pig
Grain Drying Air only	1 per bushel		
Heated	.2 per bushel	**Poultry Equipment**	
Hay Drying Air only	40 per ton	Brooding—Hover	.5 per chicken
Heated	12 per ton	Infrared	.8 per chicken
Water Pump Shallow	1 per 1000 gallon	Egg Cooler	1 per year per layer
Deep	2 per 1000 gallon	Egg Washer	1 per 2000 eggs
		Feed Grinder	5 per ton
Dairy Equipment		Feeders (auto)	.3 per year per layer
Gutter Cleaner	0.5 per month per cow	Incubator	.2 per egg
Milk Cooler can	1 per 10 gallon	Water Fountain	.1 per month per bird
Ventilator	3 per month per cow	Water Warmer	4 per month per 100 birds
Water Heater	7 per month per cow	Ventilator	.4 per month per bird
Water Fountain (auto)	1 per month per cow		

Water Requirements for the Home and Farmstead

Water for:	Amount Needed
Each member of family (all purposes)	50 gallons per day
Each horse	12 gallons per day
Each steer or dry cow	12 gallons per day
Each milk producing cow	20 gallons per day
Flushing stables and washing dairy equipment	20 gallons per day
Each hog	4 gallons per day
Each sheep	2 gallons per day
100 chickens	6 gallons per day
Sprinkling lawn or garden—1" depth per 100 square feet	62 gallons per day
½" hose with nozzle	200 gallons per hour
¾" hose with nozzle	300 gallons per hour

Sizes of Troughs and Downspouts for Various Roof Areas

Roof area	Trough diameter	Downspout diameter
100 to 800 square feet	4 inches	3 inches
800 to 1000 square feet	5 inches	3 inches
1000 to 1400 square feet	5 inches	4 inches
1400 to 2000 square feet	6 inches	4 inches

3. Housing for Animals

Dairy Cows

Most small farmers have only a few dairy cows, just enough to supply their own needs. In some areas farmers keep a few more cows in order to sell the surplus to a small local dairy outlet. The accompanying charts give basic space requirements and feed supply estimates.

General

Maternity Pens	12' x 12'	
Bull Pens	10' x 10' to 12' x 12'	
Calf Pens	25 square feet	
Doors	Single 3'6" to 4' wide, Double 8 feet.	
Light	Windows 3 to 3½ square feet per cow; electricity, one 60-watt bulb per 4 cows.	
Liquid Manure	2 cu. ft. per cow per day	
Milking Parlor*	Width	Length
Side-opening gate stalls		
3 stalls on one side	10'6"	27'4"
4 stalls on one side	10'6"	36'0"
2 stalls on each side	18'8"	20'0"
3 stalls on each side	18'8"	27'4"
4 stalls on each side	18'8"	36'0"
Chute or lane stalls		
2 stalls on each side	12'4"	22'8"
3 stalls on each side	12'4"	28'4"
Herringbone stalls		
4 stalls on each side	18'8"	22'8"
Lot		
Surface (paved)	35 to 100 square feet	
Shade		
(10-12 ft. high)	30 to 40 square feet	

*Check milk inspector for sanitation code requirements.

Open Lot (Free Stall)

Resting Area (Individual stalls)
Dairy cow	50 square feet
Replacement stock	30 square feet

Paved Area
Dairy cow	100 square feet
Replacement stock	40 square feet
Holding area for milking cows	15 square feet

Stall Size
Large breeds	4' x 7'-7½'
Small breeds	3½' x 6½'-7'
Heifers	3½' x 6'
Calves	2½'-3' x 5'

Feeding Space (Twice a Day Feeding)
Milking cows	24 inches per head
Replacement stock	12 inches per head

Bedding (baled straw)
Milking cow	1½ to 2 tons
Replacement stock	½ to ¾ tons

Water
Under pressure; 1 square foot open surface per 25 cows; water bowls in maternity and calf pens.
Summer — tank 15 to 25 gallons per head.

Feed Supply Estimates

	Large Breed		Small Breed	
AMOUNT	With Pasture	Dry Lot	With Pasture	Dry Lot
Cow Herd				
Dry matter for 100 pounds body weight, lbs.	3½	3½	3½	3½
Grain per pound milk, lbs.	⅓-½	⅓-½	⅓-½	⅓-½
Total grain per cow, tons	2-3	2-3	2-3	2-3
Hay, tons*	4-5	5.5-6.5	3.5-4	4.5-5.5
Replacements				
Grain (per calf or heifer), lbs.	500	500	500	500
Hay, tons*	1.9	2.7	1.3	1.9

*Substitute silage for hay at rate of 3 to 1

Other Herd Management Recommendations

Kind of breeding..................Artificial
Days dry (herd average)............50-65 days
      (individual cow)...........60 days
Weight at breeding
    Large cow (Holstein)..............750-900 lbs.
    Medium cow (Guernsey)..........700 lbs.
    Small cow (Jersey)................500-600 lbs.

Weight at calving
    Large cow (Holstein)..............1100 lbs.
    Medium cow (Guernsey)..........850-950 lbs.
    Small cow (Jersey)................800-900 lbs.
Calfhood vaccination.........4 through 8 months

Beef Cattle

The Pasture System

The pasture system is used throughout the United States as a basic method of beef production. Cow-calf herds are almost exclusively managed on a pasture system, although some interest has been shown recently in confined open lot systems for breeding herds. The pasture system has become less popular with intensified beef producers who desire the more rapid and consistent gains that can be obtained in confined lots.

The pasture system is used more commonly in range areas and in areas of lower land values. The advantage of the pasture system lies in the low capital investment required.

Shelter and Shade

Cattle need protection from extreme weather conditions. Results of research by the U.S.D.A. and state universities show there is a definite advantage in providing shelter for cattle in northern climates, and in providing shade in warmer climates.

Effects of Shelter on Performance of Feeder Calves

	Shelter	No Shelter
Avg. wt./head 12/10/62 (lbs.)	541.6	540.6
Avg. wt./head 4/22/63 (lbs.)	888.6	864.2
Avg. daily gain/head (lbs.)	2.61	2.43
Cost of feed/lb. of gain	11.4¢	13.0¢

University of Connecticut

Effects of Shelter on Performance of Yearling Steers in Two Winter Trials

	Shelter	No Shelter
No. steers	117	117
Avg. daily feed (lbs.)	29.5	29.6
Feed/lb. of gain (lbs.)	10.0	11.91
Feed cost/cwt. gain	$18.96	$21.98
Avg. daily gain (lbs.)	2.96	2.58

Iowa State University

Effects of Shelter on Performance of Yearling Steers in Two Summer Trials

	Shelter	No Shelter
No. steers	120	120
Avg. daily feed (lbs.)	25.8	25.6
Feed/lb. gain (lbs.)	9.5	10.05
Feed cost/cwt. gain	$17.55	$18.53
Avg. daily gain (lbs.)	2.74	2.36

Iowa State University

SECTION I: Planning

Swine Planning

A small homestead operation that wants to fulfill its own needs and have a small market supply should keep in mind that swine is the most efficient class of livestock for converting grain to meat products. Where good, consistent management practices are applied, swine production provides small producers and large complexes alike with an opportunity for quick development and fast turnover of a profitable, marketable product.

Farrow-to-Finish Operations

A farrow-to-finish program involves all phases of raising and selling. A sow herd is maintained and pigs are farrowed, grown, finished, and marketed.

A farrow-to-finish program can be used in small or large production operations where highly capable labor and management are available, along with efficient building and equipment facilities and a good source of on- or off-the-farm feed supply.

Feeder Pig Production

Feeder pig production operations are a good choice if feed grain supply is limited or grain price is high. This program requires highly skilled management and excellent facilities for intensified multiple farrowing operations.

Feeder pig production provides regular income distribution throughout the year, and is desirable for the producer who has a good supply of labor and limited capital. The feeder pig producer also offers excellent opportunities for future expansion to include growing-finishing operations.

Pasture System

Use of the pasture system for swine production has decreased because of high labor requirements and increased land values. But pasture raising of swine is still an effective production method and is practical for smaller swine operations where high investments in facilities are not desirable.

The disadvantages of pasture management compared to confinement systems include lower feed conversion, slower rates of gain, high land requirements, need for rotation, and less sanitation control.

To raise swine with least amount of complications and disease, these principles should be followed:

- Provide good clean pasture for growing and finishing each group of hogs.
- Rotate pastures on a three-year plan.
- Use portable shades and shelters for housing located on well-drained land.
- Provide portable waterers and feeders.
- Thoroughly scrub houses and clean feeders and waterers between groups of hogs.
- Relocate waterers, feeders and shelters on pasture fields as necessary to distribute manure and maintain sanitary conditions.
- Locate pasture fields on the prevailing down-wind side of the farmstead area.

For farrowing operations, either a central farrowing house or portable houses can be used. In most cases, the central house is preferable. It should be insulated and ventilated and installed with individual pens or stalls. This provides more convenience, sanitation, and effective management control of this critical stage of production.

Concrete Lot System

The concrete lot system is extensively used throughout the United States for growing and finishing swine. The concrete lot may be either completely covered or open with a partially covered area for shelter. In this system the covered area is open on one side to provide natural ventilation of the shelter area.

The advantages of the concrete lot system over the pasture system are:

Budget Analysis Guide to Determine Income and Costs

Receipts
1. Sales - hogs
   (weight x 98% x price/cwt)
   (2% shrink on farm weight)
2. Cull sows
   $\frac{\text{weight x price/cwt}}{100}$
3. Gross receipts
   (line 1 plus line 3)
4. Hog purchases
5. Net receipts
   (line 3 minus line 4)

Production Costs
6. Grain-corn equiv.
   (total bu x cost/bu)
7. Pig starter and grower feed
8. Protein supplement
9. Other
10. Labor (2.4 hrs/pig x rate/hr)
11. Veterinary and drugs
12. Miscellaneous expense (1.5% gross receipts)
13. Total production costs (sum of lines 6 through 12)
14. Total annual fixed cost (from fixed cost work sheet)
15. Total feed costs (lines 6 through 9)
16. Returns to investment and management (line 5 minus 13 and 14)

(1) minimum land required;
(2) less labor needed;
(3) more efficient gains;
(4) feeding system can be automated.

The major disadvantages are in the manure-handling requirements. With higher concentration of hogs in confinement, disease and stress can become significant hazards, especially in northern states. Required ventilation that will not cause drafts and dampness is a real problem.

The concrete lot confinement system can be used for growing and finishing pigs or for gestating sow-housing.

Swine Breeding Stock

	Per Sow	Per Boar
OPEN LOT		
Shelter resting area	25-30 sq. ft.	30-35 sq. ft.
Concrete lot	20 sq. ft.	30 sq. ft.
Dirt lot	100-200 sq. ft.	150-250 sq. ft.
PASTURE		
Pasture lot per acre	10-12 sows	5-10 sows
Shelter	20-25 sq. ft.	15-20 sq. ft.
FEEDING AND WATER SPACE		
Self feeder (grain or complete feed)	2-3 sows per hole	2 linear ft.
Supplement feeder	3-5 sows per hole	1 linear ft.
Water space	1 waterer per 12 sows	1 waterer per 3 boars

Estimated Feed Requirements for Swine

Annual Feed Requirement for Sows

With Silage

Corn silage	1425 lbs.
Corn grain	1575 lbs.
Alfalfa meal	730 lbs.
Protein supplement	265 lbs.

Without Silage

Corn grain	1820 lbs.
Protein supplement	580 lbs.

Feed Requirements Per Pig (40 lbs. to 210 lbs.)

Pig starter	12 lbs.
Pig grower	70 lbs.
Corn grain	490 lbs.
Protein supplement	75 lbs.

SECTION I: Planning

For farrow-to-finish operations, a central farrowing house that is insulated and ventilated is recommended.

For concrete lot systems, these suggestions should be followed:

- Locate buildings on well-drained sites.
- Leave one side of shelter area open with additional doors or side-wall panel openings for good ventilation.
- Provide supplemental heat in the floor.
- Construct floor slopes of ½ in. to ¾ in. per ft. in feeding and dunging areas, and a minimum of ¼ in. to ½ in. per ft. slope in bedded area. Finish concrete with steel trowel for smooth surface.
- Handle manure in solid or liquid form, depending on geographic area.
- Provide pens and pen space for maximum of 20 pigs per pen.

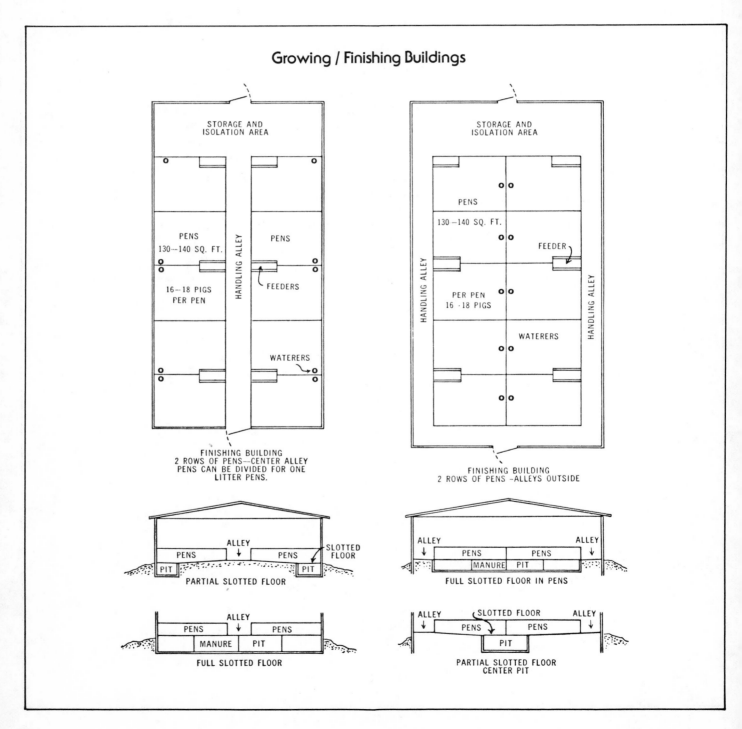

Housing for Animals

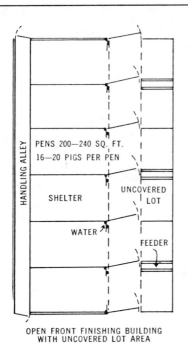
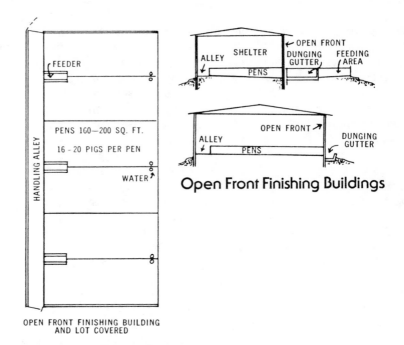

Growing and Finishing Hogs

	Weaning to 75 lbs. per hd.	75-125 lb. per hd.	125 lbs. and over per hd.
CONCRETE LOT (20-25 head per pen)			
Shelter floor area	6 sq. ft.	7 sq. ft.	8 sq. ft.
Concrete lot	8 sq. ft.	12 sq. ft.	12-15 sq. ft.
PASTURE			
Pasture per acre	20-30 hd.	15-20 hd.	10-15 hd.
Shade	5-6 sq. ft.	6-8 sq. ft.	8-12 sq. ft.
FEEDING AND WATER SPACE			
Self feeder (grain or complete feed)	6-8 hd. per hole	4-6 hd. per hole	3-5 hd. per hole
Supplement feeder	8-10 hd. per hole	8-10 hd. per hole	6-8 hd. per hole
Waterer	20-25 hd. per cup	20-25 hd. per cup	10-15 hd. per cup

Farrowing House

	Stall	Conventional Pen	Long Narrow Pen
Gilt	22" x 6' (11 sq. ft.)	6' x 8' (48 sq. ft.)	5' x 14' (70 sq. ft.)
Sow	24" x 7' (14 sq. ft.)	8' x 8' (64 sq. ft.)	5' x 14' (70 sq. ft.)
Pigs	18" each side	Creep in corner*	Creep in end of Pen*

*Both conventional pen and long narrow pen have a guard rail 8" high and 8" from pen side in addition to creep

SECTION I: Planning

Horses

Many farmers keep horses for recreational use, or transportation around the farm, or rent stalls out to others. Planning will involve a determination of space necessary for stalls, traffic lanes, storage of feed, tack, and equipment. Special-purpose features include washrack, trailer storage, breeding area, riding arena, rest space, and show stable. You should also provide for future building and paddock expansion, as well as good traffic patterns and snow removal.

Barn Styles

The trend is toward one-floor structures with clear span or post and beam supported roofs. The three shapes most widely used are the shed, gable, and offset gable.

Shed Roof. This is widely used on open-front and enclosed facilities, attached lean-to structure, and small movable buildings. It consists of a single-slope roof, and is relatively low in cost while providing good headroom. It is simple to build and insulate, and adapts well to natural lighting with spaced translucent roof panels. Most shed roofs employ a low pitch (slope) roof to keep the high side of the roof as low as possible, unless room for overhead storage is desired.

Gable Roof. The gable roof is the most widely used for both open front and closed buildings, and is applicable to narrow and wide barns. This roof costs a little more than a shed roof, and is adaptable for natural overhead lighting using fixed translucent roof panels, and for natural building ventilation through eave and ridge openings. Side extensions for roof overhang, covered path, or greater barn width can be obtained by increasing the roof height of the building without changing the pitch. The ceiling may follow the slope of the roof or be dropped to wall height, with or without overhead storage.

Offset gable. Used on both narrow and wide buildings for horse housing and related activities, the offset gable has a triangular roof with two equal slopes of different lengths that meet at a ridge.

Open-Front Buildings

When horses are kept outside they should be provided with free-choice shelter. An open-front shelter is usually sufficient. If a large building is needed, the minimum width should be at least 32 ft. wide for moderate climates and 40 ft. wide for cold climates; it also should be large enough to provide 60 to 80 sq. ft. per 1000 lb. of animal weight using the shelter. Additional room may be necessary for hay and bedding storage, pens, etc. Alleys should be 10 to 12 ft. wide; ceiling height should be at least 8 ft. for a horse and 12 ft. for a horse with a rider.

Face the open side of the building away from prevailing winds. The optimum clearance on the open side is 10 ft. Hay and bedding storage may be located inside the shelter; feed hay outside the building in racks to conserve bedding space. Provide adequate fresh water, which may mean a heated stock waterer for winter use. A single heated water bowl will serve 8 to 10 horses. Some artificial light is preferable; aim for 100 watts of light per 500 sq. ft. of floor area. Install ridge devices and adjustable panels in the back wall for summer and winter ventilation.

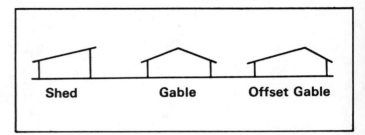

Item	Tie Stalls	Box Stalls
Water	out-of-stall	in-stall
Feed	in-stall	in-stall
Manure	less carrying	more carrying
Bedding	less required	more required
Exercise	out-of-stall	limited in-stall
Space	45 to 60 sq ft	100 to 320 sq ft
Floor	clay or plank	clay or plank
Partitions	strong & tight	strong & tight
Top Guard	manger end	on partitions

Bedding material approximate water-absorbing capacity

lbs of water/lb of bedding	
4.0	Tanning bark
3.0	Pine chips
2.5	sawdust
2.0	shavings
1.0	needles
1.5	Hardwood chips, shavings or sawdust
2.5	Shredded stover
2.1	Ground cobs
2.8	Oats, threshed
2.5	combined
2.4	chopped
3.0	HAY, chopped mature

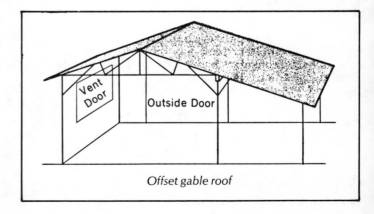

Offset gable roof

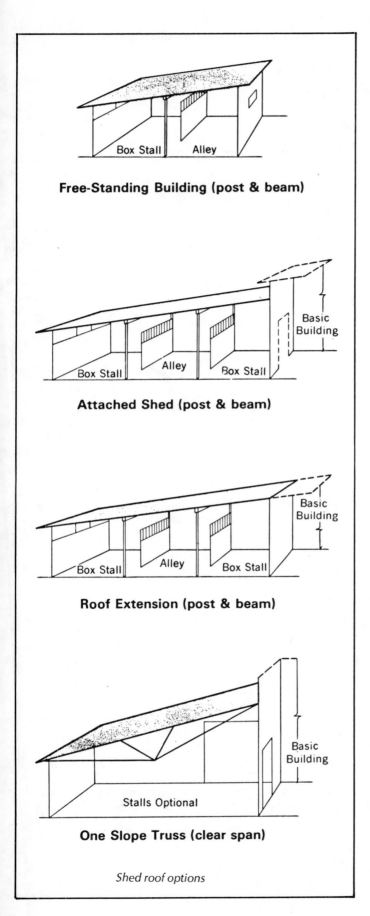

Shed roof options

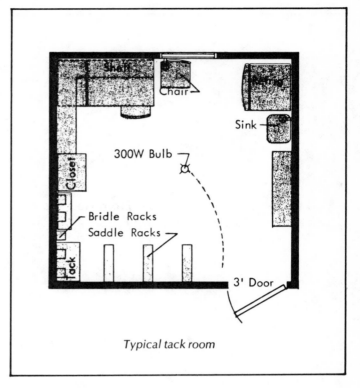

Typical tack room

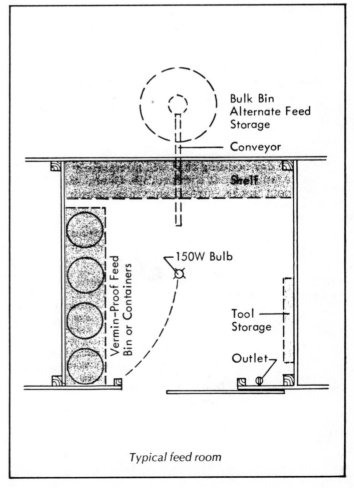

Typical feed room

SECTION I: Planning

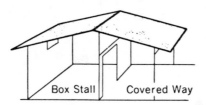

One Row Of Stalls Serviced From Covered Way
(post-beam-wall)

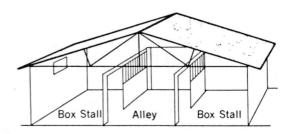

Two Rows Of Stalls Serviced From Center Alley
(clear span truss)

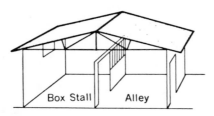

One Row Of Stalls Serviced From Enclosed Alley
(clear span truss)

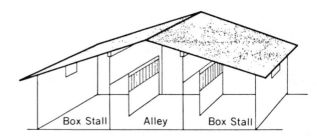

Two Rows Of Stalls Serviced From Center Alley
(post & beam)

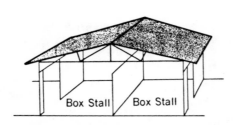

Two Rows Of Stalls Serviced From Outside
(clear span truss)

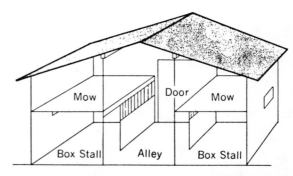

**Two Rows Of Stalls Serviced From Center Alley
Mow Over Stalls (post & beam)**

Gable roof options

Space Requirements for Horses in Buildings

	Dimensions of Stalls including Manger	
	Box Stall Size	Tie Stall Size
Mature Animal (Mare or Gelding)	10' x 10' . . . . small	
	10' x 12' . . . . medium	5' x 9'
	12' x 12' . . . . large	5' x 12'
Brood Mare	12' x 12' . . . . or larger	
Foal to 2-year old	10' x 10' . . . . . . average	4½' x 9'
	12' x 12' . . . . . . large	5' x 9'

Stallion[1]	14' x 14' . . . . . . or larger		
Pony	9' x 9' . . . . . . average		3' x 6'

[1] Work stallions daily or provide a 2-4 acre paddock for exercise.

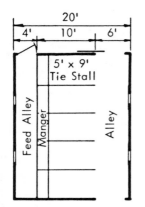

One Row, Feed Alley

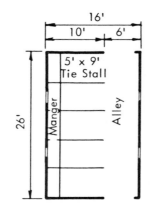

One Row, No Feed Alley

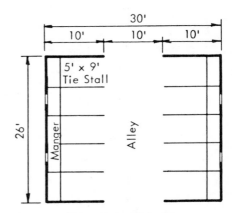

Two Face-Out Rows, No Feed Alleys

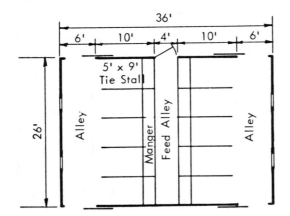

Two Face-In Rows, Feed Alley

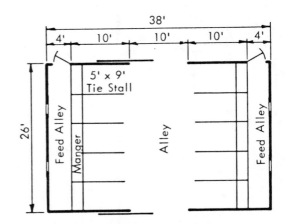

Two Face-Out Rows, Two Feed Alleys

Layouts for Tie Stalls

SECTION I: Planning

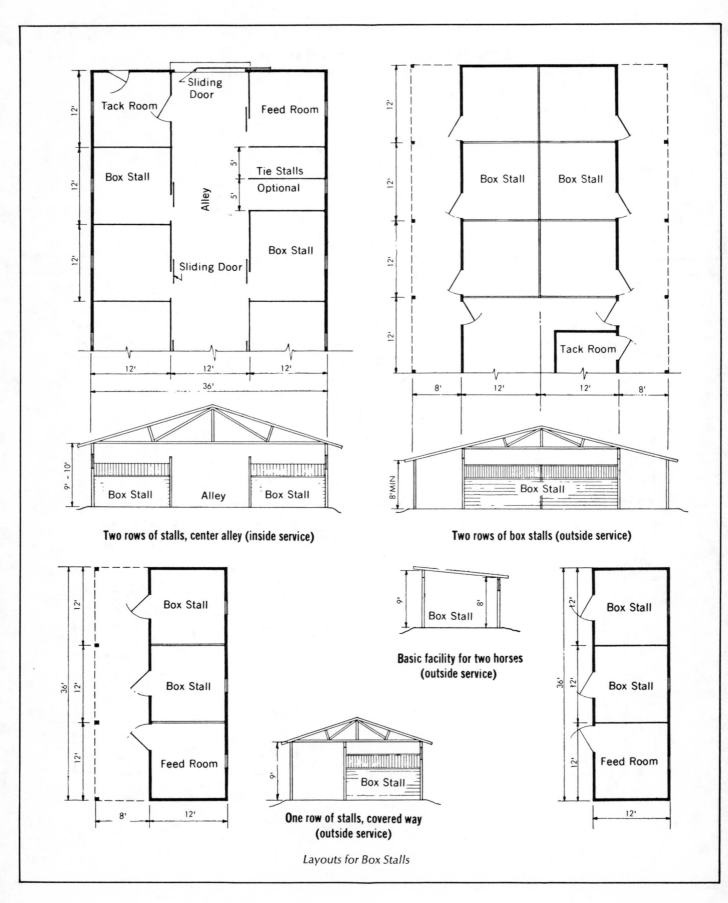

Layouts for Box Stalls

Poultry

In addition to the basic requirement of genetically well-bred birds capable of a high rate of reproduction, success in raising poultry (layers, pullets and broilers) depends on proper feeding, sound management, and good sanitation. The first area does not really fall under the scope of this book; for information write to your Cooperative Extension in Agriculture.

Chicks

Before bringing in chicks, clean the building thoroughly. Apply an approved insecticide if lice, mites, beetles or other insects have been a problem. Rinse equipment with a disinfectant. Put clean, dry litter in the building: chopped straw, wood shavings, or crushed or coarsely ground corn cobs make good litter.

Brooder Stoves. The number one requirement is heat. Brooder stoves use either gas or oil, and have hovers that contain or retain the heat close to the floor. Electric heat lamps (without hovers) are also used to brood chicks. Put the stove into operation at least 24 hours before chicks arrive in order to make adjustments; provide a temperature of 90 to 95° F. Use only new or cleaned chick guards at least 3 to 4 ft. from the edge of the brooder hover for the first 7 days. Use solid chick guards in cold houses. Reduce temperature under the hover over a period of several weeks. The chicks will gradually learn to regulate their location within the temperature zone most comfortable for them. You will gradually raise the brooder hover or heat lamp over a period of several weeks until the chicks become accustomed to the fact that the heat source is not necessary. If chicks are too cold, they will chirp; if too warm, they will lay down or pile in corners. When comfortable, they will form a ring on the floor under the heat source.

Chick Guard. This is a paper, cardboard, or fine-wire mesh ring around the heat source that confines the birds to the warmed area. It should be 12 to 18 in. high.

Housing Requirements (Birds of All Ages)

Ventilation. Adequate ventilation is essential; it helps remove ammonia odors, excess moisture, and airborne disease organisms. A small fan with a thermostat is useful for small houses. Do not protect birds too much, but avoid letting water freeze. Electric heat tapes on the water pipes and a heat bulb over the waterers will help keep water fluid in cold temperatures. Also remember that birds do not thrive in drafty areas.

Insulation. Birds produce heat; the insulation will confine this heat to the house during cold weather in larger houses. However, a small flock will not provide enough body heat to maintain a house temperature of 45 to 65° F. In most cases, 2 to 3½ in. of fiberglass batts or the equivalent will provide adequate insulation

Space Requirements. Allow ½ sq. ft. of floor space per chick up to 3 weeks of age, 1½ sq. ft. up to 14 weeks of age, and 1½–2 sq. ft. per adult. Allow 40 linear feet of feeder space per 100 hens. You should have enough feeder space so all birds (of any age) can eat at the same time. The feeder stand should be raised as the birds get older, at least to the level of the bird's back as it stands on the floor. The birds should have to reach up and over the edge of the feeder; this prevents feed waste. Allow three ½ gal. water founts per 100 chicks at one day of age; after a week, use larger founts so the water intake is unrestricted. Clean the founts every day. Allow one 5-gal. water fountain or 4 linear ft. of watering trough for each 100 hens. Water is the cheapest nutrient, but it must be kept clean, fresh, and available.

Lighting. Light is necessary for laying hens. It stimulates egg production; hens will not lay well if kept under conditions of declining light-day length, which occurs from late June until late December. Pullets (young female chickens) become sexually mature and will lay eggs at approximately 20 weeks old. They should be given 16 hours of light (daylight plus electric) per day. The amount of light required is not high. One footcandle of light at bird level in the dark part of the house is adequate; a 40-watt bulb each 100 sq. ft. of floor space is more than sufficient.

Cannibalism

Not meeting the requirements for heat, space, feed, and light can result in cannibalism. Measures to avoid this condition include varying temperatures and light within the house when possible, and allowing the birds some light and some dark areas. If possible, allow birds to run free part of the time. If these controls are not sufficient, debeaking is the only sure cure.

Sanitation

The first step is a thorough cleaning and elimination of insects and mites before housing a new flock in the fall. Ongoing measures include:

(1) removing manure from beneath caged layers as frequently as feasible;
(2) scattering manure lightly outdoors so that fly eggs and larvae can be killed by drying (avoid piling manure, or you will create a fly breeding site);
(3) removing floor litter, droppings, and nesting materials at regular intervals;
(4) maintaining low moisture levels in the manure with proper ventilation and quick repair or replacement of leaking water troughs, individual water valves, or other faulty fittings.

Mechanical Systems. Use a fan to blow air through a screened doorway from the egg room or other work areas into the main poultry house. Flies will not move against the wind into the egg room or other areas.

Baits. Never use baits where loose birds are housed. Scatter bait where flies rest on floor areas not satisfactorily treated by sprays.

Atomizers. Use pyrethrum oil-base sprays (0.1% pyrethrins plus 1% piperonyl butoxide) at the rate of ½ fluid oz. per 1,000 cu. ft. on a daily basis for fly control in closed egg rooms.

Portable Mist Machines. Mechanical foggers are labor-saving, particularly for caged birds. Follow label instructions and precautions.

Fly Traps and Electrical Grids. These often trap and kill large numbers of flies, but they are not sufficient to completely control fly populations. They should be used as a supplement to insecticide control.

Floor System

The floor system has long been used for layer flocks and is still popular with smaller producers. With partial or fully-slotted floors and automated feeding, it can be highly efficient for egg production operations. The floor system will continue to be popular for the smaller flock operations under 5,000 birds and with heavy breeder layer operations.

In planning and developing floor systems for layer flocks, the following recommendations should be considered.

- Locate buildings on well-drained sites away from frequently used drive lanes or roads, and a minimum of 300 ft. from the farm residence.
- Install slotted floors with a minimum of two-thirds slotted and one-third litter floor.
- Place feeders and waterers on slotted floor areas.
- Install nest banks in line for convenient collection.
- In controlled-environment houses provide effective insulation and ventilation system for a greater density of birds.
- In an open house provide panel openings for a maximum of two-thirds of the wall area. Provide ridge ventilation and ceiling insulation. Use white color-coated galvanized steel roofing for heat reflection; use white or other shades of color-coated galvanized steel siding.

Duck and Geese Production

Duck and goose meat and egg production have not followed the same dramatic growth in the poultry industry as chickens and turkeys.

Egg demand and production has declined and marketing has been confined to selected market outlets.

Meat production has increased steadily, but specialization in production and processing has been slow to develop due, probably, to the continued seasonal demand.

Small farm production flocks are still prevalent, particularly with geese; however, more commercial farm operations with intensified production are developing and will continue to supply a greater share of the market demand in the future.

Breeder flock operations for producing hatching eggs for duck and geese meet production, and raising ducks and geese for market on a commercial production scale, provide profitable opportunities for producers.

Breeder flock and rearing operations are profitable in small or large commercial operations, but because of seasonal demand operations are best suited to farm programs where seasonal labor is available.

Duck Production Housing

Rearing ducks requires good building facilities to keep ducklings warm, dry and free of drafts. Supplemental heat is required for about 4 weeks. Ducklings with a good start at brooding grow rapidly and are efficient feed converters; they finish for market at 8 to 9 weeks of age.

Following are recommendations for housing management for rearing ducks:

- Provide a warm, dry well-ventilated house for brooding and growing.
- Divide house into pens with or without adjoining outside lots and rear ducks in flocks of 500 or less per pen.
- Use wire or slotted floors in house. Litter floor for small flocks is practical but must be kept dry. Where litter floor is used, place waterers on wire platforms over a drain.
- For confinement from rearing to finish, provide house floor space of 2.5 sq. ft. per duckling. Where outside lot is used, provide building space of 1 sq. ft. of house floor space and 70 sq. ft. of lot space per duckling.
- Where lot is used, locate watering facilities at the far end or lowest point of the lot.
- Swimming water is not necessary but many growers provide it after ducks are 5 weeks old. Swimming water can be provided by a stream or pond or by building an artificial pool.

Ducks are adaptable to a wide range of environmental conditions. Major needs are for shelter from weather extremes and protection from predators.

Breeder flocks need very limited facilities. A shelter building, with three sides closed and an open front with adjoining dirt lots, will provide good facilities for managing the flock.

During laying periods breeders should be confined to the house at night and have access to lots during the day.

Following are recommendations for housing the breeders.

- Locate buildings and lots on well-drained soils.
- Provide a well-ventilated shelter building with adjoining sloped lots. Divide space into pen area to handle 200 to 250 breeders per pen.
- House floor should be kept dry with the change or addition of litter and good ventilation. Concrete floor is preferable unless a well-drained soil condition exists.
- Provide water and feeding facilities in the lot area.
- Provide 5 sq. ft. of floor space per bird in the house, and a minimum of 35 sq. ft. per bird in the lot.

- Swimming water is not necessary but can be provided by concrete or steel troughs.

Geese Production Housing

Geese are relatively easy to rear and manage and have very limited requirements for housing and other facilities.

Breeder flocks can be maintained in good condition on good grass or legume pastures. An open shed provided for shelter of the flock during weather extremes is all that is required. Nests should be placed on the floor along the walls for use during the laying season.

A shelter building placed in the center of a field area, with the field fenced into 3 to 4 pasture lots on which the flock can be rotated, is ideal for breeder flock management.

Rearing goslings for meat requires a good brooding building and facilities to get the goslings off to a good start. Supplemental heat is required for about 4 weeks during the brooding period. Goslings should be given access to succulent pasture as soon as possible after the brooding period.

Geese are marketed either at an early finish of 10 to 12 weeks of age, at 7 to 8 pounds, or are held and finished for the Christmas/New Year market at 10 to 15 pounds (live weight).

Confined open lots, well shaded or with shelter, should be provided for finishing of birds.

SECTION 2: Setting Up the Basic Farm Structure

4. Money Matters

Annual Costs

Annual costs of owning your new or remodeled buildings should be balanced against the benefits you expect to get from them.

We have stressed in this book that your building should be a profit-making and useful tool for you. But you have to decide just how profitable it will be. To do that, figure your total yearly costs by adding up the following factors.

Depreciation

This figure represents the loss in value your builidng will suffer each year. How much depreciation you can charge off depends upon the latest tax ratings, but a commonly accepted schedule is 10 percent of the building's value each year for temporary or movable buildings, 5 percent for open-type structures, and 2½ percent for well-constructed, typical wood-framed buildings.

These depreciation rates apply only to new structures. If you decide to remodel, then you can depreciate in a shorter period of time (at a higher percentage of the value each year).

Financing Costs

Interest on the money you used to build the building is an operating cost. You can charge off to expenses the full interest on the money borrowed to build or remodel.

Maintenance

Repairs, taxes, insurance, and incidentals usually run about 3 percent of the building's cost per year. However, you should keep accurate records to determine this.

Construction Plans

At this point, you will have a general idea as to the kind of building you need. To determine the specific facilities you want in the building, read through Chapter 5, "Building Basics". Now you are about ready to think about the actual construction of the building itself and how it is to be done.

If you have decided that a prefabricated building, such as a standard brooder, hog house, grain bin, or similar structure will fill your needs, there is no problem. In many cases, the total cost of a prefabricated building will include the cost of erection on your site. Or you will be quoted an erection cost, if it is offered separately. But if you need a large custom building—that is, a building that is tailored specifically to your needs and operation—you have a different kind of worry. You will have to make up your mind whether to build it yourself or to hire a professional farm building contractor, such as a member of the National Farm Builders Association (NFBA), to get the job done.

What About Using Subcontractors?

You may want to make use of subcontractors to get special jobs done. For example, if site preparation requires a considerable amount of earthmoving, you might want to contract with a bulldozer operator to handle it. Or you might call in a concrete contractor to lay a large or difficult floor. This will get you professional help on jobs that are beyond your scope; however, there are some built-in difficulties. You will have to make all the arrangements yourself, coordinate their activities, and supervise each job to be sure it gets done the way you want it. And you may find that a particular subcontractor is not available at the time you need him.

After you add up all these factors, you may want to consider the benefits of having your building erected by a professional farm building contractor. Many of the best are members of the NFBA, so you know they are farm building specialists.

Using a Professional Farm Building Contractor

The professional has an engineering staff who spend most of their time on farm building design. They have prepared building plans for many farmers with requirements similar to yours, so they bring a lot of experience to the discussion of your particular needs. When your plan is completed, it will incorporate the latest techniques in farm building erection as well as the newest and best materials. So your building can be built at minimum cost, and built to last.

The professional farm builder is aware of the environmental requirements of your location, so the building will be designed with the right roof, and snow and wind loading factors. He will furnish blueprints and other information needed to get county and state permits and compliances. He will counsel you on questions such as the value of pre-painted roofing and siding versus unpainted...or the relative merits of various insulating materials...or the types and kinds of doors, windows, and fixtures that you might want.

The professional builder also provides a guarantee on both materials and services. That means that if there is a building failure due to faulty materials or poor construction methods, the contractor will make good.

He also carries insurance against all liabilities that could arise during construction. If someone falls off a ladder or has a panel fall on him, the contractor's insurance covers these eventualities.

For the less complicated types of construction, the savings you can gain by doing it yourself are worth it; for the larger projects a contractor should be seriously considered unless you have had prior construction experience.

Now that you have looked at all the possibilities, you will have come to a decision. If you have determined that your best bet is a professional farm building contractor, but you are not able to find one in your area, write the Farm Roofing & Siding Sales Department (W6L0, Wheeling Corrugating Company, 1134 Market Street, Wheeling, West Virginia 26003), and tell them what you need. Or write to the National Frame Builders Association, (1406 Third National Building, Dayton, Ohio 45402).

5. Building Basics

Site Selection and Use

One basic site requirement is that drainage from feedlots must not go toward the well or into a stream or roadside ditch and cause pollution. It is also desirable to have a clear view from the house to the drive or road. Listed below are other major considerations.

Soil Type. Porous, loose soil dries fastest and makes a good solid base for foundations and floors.

Wind. Protect your building from winter winds and expose it to summer winds to increase both your and your animals' comfort in the building. In most geographic areas, buildings should be open to the south or southeast for protection from prevailing winter winds. To increase ventilation in very hot areas, it may be important to place the building so it will "catch" the prevailing summer winds. Livestock buildings should be placed so that prevailing summer winds do not carry odors to area homes.

Sun. In cold regions, run the long side of the enclosed buildings north and south to get the most heat from the morning and afternoon sun. In warm regions, your best bet might be an east-west orientation to reduce cooling and ventilation problems.

Fire Hazards. You are sure to create a fire hazard if you place your buildings close enough so that fire can spread easily from one building to another. But placing them too far apart reduces your efficiency. To balance travel distances and labor efficiency against fire risks, leave at least 75 ft. between buildings. This distance should not seriously cut your operating efficiency, but will allow enough room to position firefighting equipment between your buildings and prevent fire from jumping building to building.

Utilities. Water, gas, electricity, and telephone connections should be located nearby. They should be able to handle the increased load your building will place on them. If you have any doubts about the capacities of any of your utility services, check with your local utility company.

Architecture. Your building should not only have a useful and efficient layout, it should also be attractive. Its exterior should be a pleasing color and it should blend in with your other buildings. If you place your buildings at right angles to each other, they will be more pleasing, for example, than buildings that are placed haphazardly or in a line. A word of caution: Don't let neighborhood architecture influence you. The attractive building on a neighboring farm might fit your neighbor's needs, but that doesn't necessarily mean it will fit yours.

Temperature and Humidity Variation. Average, high, and low temperatures in your area can affect your choice of building foundations, floors, insulation, and ventilation. All these points will be discussed in later chapters.

Site Preparation and Building Layout

After you have picked your site, and the building's size, configuration, and orientation, you should carefully consider site preparation and building layout before starting the actual construction.

First, clear the site of trees, rocks, and other debris that might interfere with building. Do whatever grading or earth-moving necessary to make the best site for your needs.

Second, stake out the building on the site so that it is square with other buildings and with itself. To do this, make up a right (90°) triangle out of boards, with dimensions of 6 ft. by 8 ft. by 10 ft.

Batter Boards

Once you have placed stakes at corners and midpoints using your triangle to make sure the building is square, you will be able to place batter boards. The boards are nailed to stakes and act as fastenings when stretching mason's line cords between board pairs. The boards will give proper alignment and height for the footings and foundation.

The boards themselves need not be in perfect alignment but their top edges must be exactly level. This means nailing up the boards by means of a level-transit so that each board's top edge is a uniform height. It is best to rent a level-transit, unless you will be doing so much construction work that buying one wold be economically worthwhile. "The instrument" (as it is called by construction men) screws to a tripod and the mounting includes adjustment devices for leveling and sighting; in this way it remains level when rotated to a sight in any direction. The tripod is basically a telescope; it has vertical and horizontal cross-hairs in the lens and an adjusting dial that permits focusing for various distances.

The batter boards should be 1 x 6's or 2 x 4's; they will be nailed to ground stakes slightly above the ground in horizontal position. They are placed at building corners and inter-

Building Basics

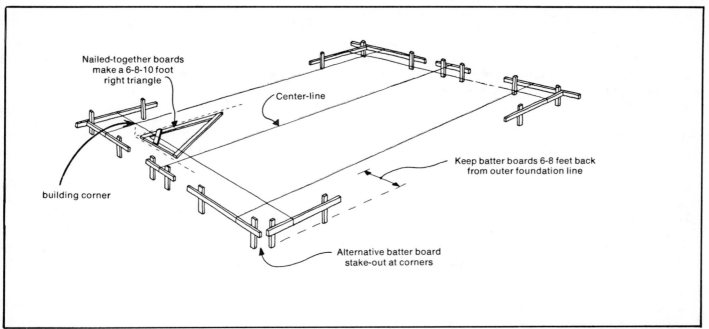

Mason's cord, stretched from level batter boards at markings indicating the width and length of the building, will cross at building corners. (Art reproduced from "How to Build Your Own Home," a Successful book by Robert C. Reschke)

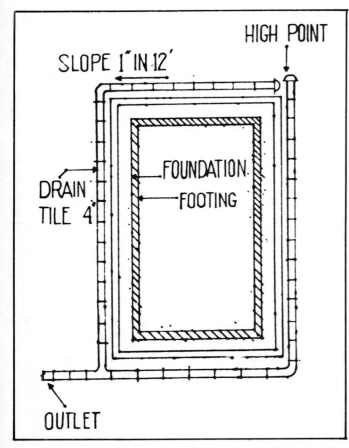

Drainage tile or perforated conduit is placed in trench excavated to just below the bottom of the footing at its highest point, shown upper right corner. (Art courtesy of Wheeling-Pittsburgh Steel Co.)

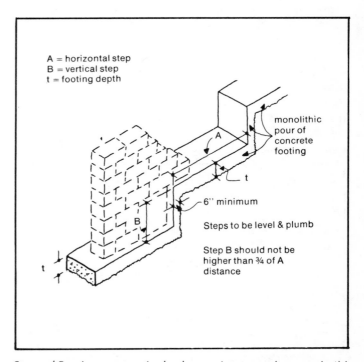

Stepped Footings are required to be continuous under some building codes; in such cases, the earth base must be dug to grade on the different levels with vertical sections spaded smoothly. Form board or plywood is nailed to long stakes to form the vertical wall face for the footing and this form must be well-braced to withstand the concrete pressure during pouring. (Art reproduced from "How to Build Your Own Home," a Successful Book by Robert C. Reschke.)

mediate points. When mason's line is stretched between marks on the batter boards, the line marks the proper position of the outside face of the foundation wall. When a set of batter boards are placed so their top edges are all horizontal and at the same height, then the stretched lines can also become reference points for determining proper and accurate heights of footings and foundation walls (if the construction project you have chosen requires them).

Now you are ready to grade the site to meet your floor condition. If the site is smooth and you are going to put up a building with an earth floor, then your grading may only require smoothing out some bothersome lumps. But if your site is rocky or somewhat hilly, then it is best to make it as smooth as possible before you start to build. Or, if you are planning a concrete or wood floor, you should grade the site first so you will not have to do the work in an enclosed space later.

Final elevation of the ground should also be considered. If you have to grade to get a smooth floor, then you must be careful not to get the floor too low for good drainage. If you are going to put in a concrete floor, you should grade to about one foot below the finished floor elevation to allow for porous fill and concrete thickness.

Foundations

Foundations support your building by spreading weight over a large area of subsoil. They are needed for conventional lumber or steel-framed buildings, but are not needed for square post buildings except when they must support heavy loads.

Foundations are usually concrete or masonry footings topped with walls or piers. They can run continuously around the building's outside wall, or can be placed under the building's columns only. The most common farm building foundation runs continuously around the building because it is easier for you to attach the sill to a continuous and smooth surface. Also, the sill can be lighter, and will not sag if it is bolted to a continuous wall.

One of the most important and least understood functions of a foundation is its ability to allow your building to settle evenly. You can expect your building to settle somewhat, but it will not be a problem if the rate is the same for all of the foundation. But when one part of your building settles more rapidly than another, the building "racks." When this happens the frame can be seriously pulled out of shape. The frame's joints will often fail, and wall and roofing sheets will often warp and crack to the point where siding and roofing nails pop out. Then you have a repair job on your hands, one that you can avoid with proper planning, design, and construction.

Your first foundation planning step is to find out what type of soil you are going to build on. Some soils can support more weight than others, which will affect footing size and settlement rates. For comparison purposes, the accompanying table lists seven common soil types by their ability to carry weight, with the least desirable first.

The table also shows that one square foot of hardrock can hold fifteen times the weight of a square foot of soft clay. This means that footings on soft clay must have fifteen times the area of footings on hard rock if they are to settle equally.

The Suggested Footing Sizes Table shows suggested footing sizes for various types of farm buildings placed on ordi-

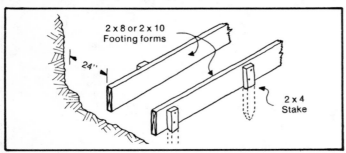

Footing forms are set parallel with top edges level and at uniform depth measured down from line cord stretched between batter board pairs. Actual depth will depend upon height of foundation wall. Where footings of different levels meet, the common practice is to end the footing at each level with ends as nearly vertical one over the other as the soil condition will permit. At end points, short planks should be nailed to the parallel side forms. If form planks are set so their bottom edges are a half inch or more above grade, avoid loss of concrete by placing earth fill on outer sides of forms to keep the concrete from flowing out. (Art reproduced from "How to Build Your Own Home," a Successful Book by Robert C. Reschke.)

Soil Weight-Carrying Capacity	
Type of soil	Relative weight-carrying capacity
Soft clay, sandy loam, or silt	1
Ordinary clay or sand	2
Moderately dry clay or fine sand and dry sand	3
Hard, dry clay or coarse, firm sand	4
Gravel	6
Soft rock	8
Hard rock	15

Suggested Footing Sizes					
	Continuous foundation footing (in.)		Separate column footing (in.)		
Building	Width	Thickness	Length	Width	Thickness
One-story poultry, hog house, machine shed	16	8	18–30	18–30	9–15
Barns, larger granaries, corn-crib, 2-story poultry or hog house	24	12	24–36	24–36	12–18

nary clay or sand. To use this table with the table above, simply double the footing width if your soil is soft clay, sandy loam, or silt, and halve the footing's width if your soil is hard, dry clay or coarse, firm sand.

If the soil type changes from one end of your building to another, it is safest to pick the worst type of soil and size your footings to suit it. A more economical practice, however, is to adjust the size to meet the soil conditions. By adjusting the footing size your settlement rate will be more even.

But footing size is not your only consideration; you also must consider how deep underground to place the footing. And depth depends on the area climate, since footings must be placed below frost. If they are not, the footing—and your building—will rise when the ground freezes and fall when it thaws, often unevenly, and often causing the foundation to tip. This uneven movement will almost always lead to severe frame racking. *(See Appendix A.1 for depths.)*

The most common foundation materials include concrete, brick, block, and stone. We recommend that you use concrete whenever possible, because when placed it molds itself perfectly to the subsoil so that it supports weight better than other materials.

If you use brick, pick a hard-burned type that has a low porosity. If block is used, be sure to choose a concrete (not cinder) block for greater strength and lower porosity. And be sure to fill the cores on all underground block with a high-quality mortar.

While it is best to make your footings from concrete, foundation walls are usually block. But they can also be cast-in-place concrete, brick, or stone. How thick you make your foundation wall will depend upon the material selected and the type of building you're planning to build.

It is good practice to use reinforcing steel bars in farm building foundations. For one-story buildings with only vertical loads on the foundations, two bars in the footing and two bars in the top of the foundation wall are usually enough. However, for tall heavy buildings, an engineered design should be used.

You will need some type of foundation drainage unless your building site is exceptionally well drained and slopes away from the building in all directions. It is important that underground water is taken away from the foundation quickly. If not, it could freeze and thaw, and eventually seep through foundation walls and cause serious building damage.

If your site's water conditions are not too severe, you will probably get enough drainage by simply back-filling the outside of the foundation with coarse gravel or crushed rock. Start the gravel at the base of the footing and fill up to at least 12 in. over the top of the footing.

Where conditions are more severe, it is wise to place a tile or perforated conduit completely around the building wall. Use a 4-in. diameter tile or perforated conduit. Place it in a trench excavated to just below the bottom of the footing at its highest point. Slope the tile 1 in. per 12 ft. to an outlet.

Then backfill the tile with 12 in. or more of crushed stone or coarse gravel placed over and around the tile and against the footing and wall.

Then you can finish backfilling with a layer of clay followed by a layer of topsoil over the porous gravel.

Concrete—Quality is Important

Concrete is a mixture of portland cement, water, and aggregates. Aggregates—sand (up to ¼ in. in size) and gravel or crushed stone (¼ inch to 1½ inches in size)—should be clean and hard. Water should be actually fit to drink. The ratio of cement, sand, stone, and water is called the mix.

The cement and water form a paste that coats all the aggregates and binds them together. The consistency of the paste greatly affects concrete quality: a rich, thick paste will produce concrete with greater strength than will a thin, watery paste.

Paste consistency depends upon the amount of water added to the concrete mix per sack of cement. This water-cement ratio is commonly expressed as the number of gallons of water per sack of cement. For example, a "5-gallon mix" specifies 5 gallons of water per sack of cement.

Coarse aggregates are added to the concrete mix because they are more economical than sand; it takes less cement to coat a 1-in. stone than it does to coat enough sand particles to equal the same volume as the stone.

The maximum size of coarse aggregates is important. A good rule of thumb is to never use aggregates bigger than one-fifth the thickness of vertically formed concrete, and one-third the thickness of flat slabs. For reinforced concrete, aggregate should not be bigger than three-fourths the distance between bars.

The workability of the mix that you want determines its proportion of sand to coarse aggregate. More large aggregate can be used in thick forms, such as footers, than in thinner floor slabs or vertical walls.

Shown are mixes for various types of work. When using the table, first select the gallons of water per sack of cement. Keep this constant and adjust the sand and coarse aggregate until the workability is right.

Air-entrained cement should be used for all exposed farm concrete. It virtually eliminates scaling due to freezing, thawing, and salt action.

Ready-mix concrete can be ordered by the water-cement ratio, but more often it is ordered by specifying sacks of cement per cubic yard, air entraining, and maximum size of aggregate. The mix will be delivered with a workable proportion of sand and coarse aggregate; it should be mushy, but not watery enough to flow.

Placing and Curing

The concrete should be in place within an hour after beginning to mix or before it begins to set. Make sure forms are strong enough. When placing concrete in vertical forms, it is

Concrete Mixes

Kind of work	Max. size aggregate (inches)	Gal. of water added for each sack of cement, using:			Suggested mixture for 1-sack trial batches[4]			Ready-mix sacks cement per yard[5]
		Damp[1] sand	Wet[2] (average) sand	Very Wet[3] sand	Cement, sacks (cu. ft.)	Aggregates		
						Fine, (cu. ft.)	Coarse, (cu. ft.)	
"5-gallon mix" Concrete subjected to severe wear, weather, or weak acid and alkali solutions	¾	4½	4	3½	1	2	2¼	7¾
"6-gallon mix" Floors (home, barn), driveways, walks, septic tanks, storage tanks, structural concrete	1	5½	5	4½	1	2¼	3	6¼
	1½	5½	5	4½	1	2½	3½	6
"7-gallon mix" Foundation walls, footings, mass concrete, etc.	1½	6¼	5½	4¾	1	3	4	5

[1]Damp sand will fall apart after being squeezed in the palm of the hand.
[2]Wet sand will ball in the hand when squeezed, but leaves no moisture on the palm.
[3]Very wet sand has been recently rained on or pumped.
[4]Mix proportions will vary slightly depending on gradation of aggregates.
[5]Medium consistency (3-in. slump).

necessary to spade the mix next to the forms to eliminate honeycomb. When placing flat slabs, the top of the concrete is struck off with a straight piece of framing lumber by seesawing it back and forth. Then the surface is floated to a flat but gritty surface with a large wood or magnesium float. Magnesium floats are better for air-entrained concrete.

For sidewalks and yard pavements, this gritty surface can be the final finish. If a very smooth surface is needed, a steel trowel is used after the concrete has become quite firm but not dry. A steel trowel should not be used on freshly placed soft concrete, since it sucks the cement to the surface. This can cause checking and spalling.

Concrete must be kept damp for at least five days to properly cure it and let it gain maximum strength. A common curing procedure is to wet the newly set concrete and cover it with polyethylene, burlap, or straw. Another method uses a light, continuous sprinkling system.

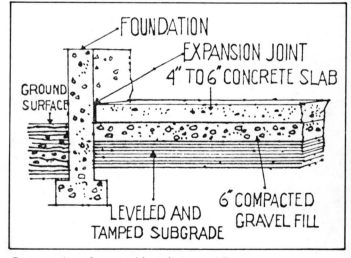

Cross-section of a typical foundation and floor.

Cold Weather Concrete

More caution is needed when placing concrete in freezing weather. Never place concrete on frozen ground; later thawing will cause it to settle and crack.

The mix temperature should be between 50 °F. and 70 °F. Heat the water up to 180 °F. and, if necessary, also heat the aggregates prior to mixing to assure that the necessary mix temperature is maintained.

Two pounds of calcium chloride per sack of cement can be added to the water. Calcium chloride is not an antifreeze but it will make the concrete set faster and enable you to cover it before it can freeze. If you use calcium chloride, keep the concrete at 50 °F. for four days; if calcium chloride is not used, the concrete should be kept at 50 °F. for seven days.

Indoors, heaters can be used to keep the concrete warm. But if the heater is not vented, fumes will react with the concrete and cause a dusty surface.

General Considerations

1. Clean the reinforcing steel before placement; otherwise oil and mill scale will prevent the concrete and steel from bonding properly.

2. Foundation walls should extend 12 to 18 in. above ground level. This will hold the sill above ground moisture and discourage termites. For additional termite protection, add a termite shield between the foundation and sill.

3. Center foundation walls on footings whenever possible. Placing the wall on one edge of the footing will reduce the footing's ability to spread the building's weight evenly to the ground.

4. Foundation drainage is important. Unless you have an exceptionally well-drained site, you should at least backfill footings with gravel and possibly should install tile drains.

Floors

Because of the trend to one-story buildings, floors in farm buildings are usually placed on the ground or supported just above ground level. Depending on their intended use, floors can be simply packed dirt or complicated slatted structures; they can be made of earth, concrete, wood, wire, structural steel, or other metal shapes. In general, you should plan your floor to cover these potential problems.

Support for intended load. It doesn't make economic sense to put in a floor that is too strong or too weak, so build it to meet the conditions under which it will function.

Easy cleaning. Always try to slope floors toward an opening or drain for easy and quick hose-down; and if your floor is concrete, make it smooth enough so water will flow easily and without leaving puddles.

Nonslip surface. Since you can't have a truly nonslip surface that is also perfectly smooth and easily cleaned, you will have to weigh the desirable features of cleanability against safety and arrive at a compromise. Your concrete floor, for example, can have a reasonably nonskid surface and still be easily cleaned if the surface is wood-floated, broom-finished, or acid-etched.

Resist chemical action. Your floor should be able to resist farm acids, caustic alkalies, and other chemicals that can rapidly deteriorate concrete, wood, and steel. For example, stored fertilizer will often combine with water to produce sulfate acids. Therefore, concrete floors in fertilizer-storage areas should be made of special high-alumina concrete, and extra waterproofing precautions should be used.

Resist wear. Your concrete floors will resist abrasion for a much longer time if you use surface hardeners such as iron filings, carborundum, granite chips, or commercially available liquid hardeners.

Concrete Floors

Concrete is by far the most popular flooring material. Concrete floors are easily placed, are long lasting, can be made to resist chemical action of all types, and their surface

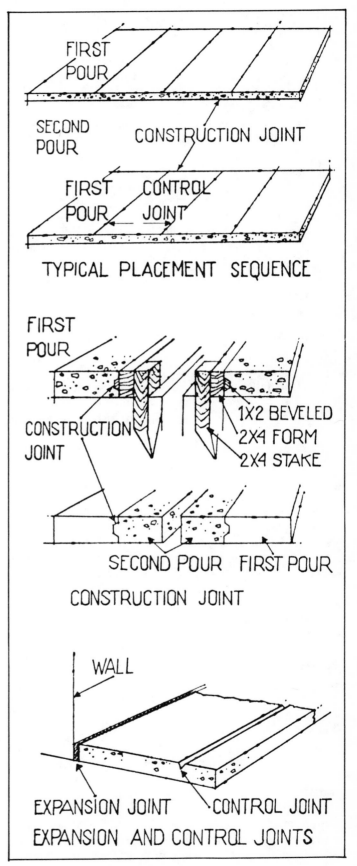

Usual placement sequence of a concrete floor.

SECTION II: Setting Up the Basic Farm Structure

can be nonskid and easily cleaned—if you follow these simple guidelines.

- Be sure the soil under the concrete slab has been compacted, preferably with a roller or mechanical tamper. If you place concrete over loose soil, your floor will settle and crack.
- Reinforcing mesh placed in a slab will help to control cracking if properly located in the center of the slab. It does not prevent cracking due to settlement or shrinkage in large slabs, but it will keep these cracks small.
- To ensure a dry floor, place the slab on a membrane of polyethylene sheet laid on 6 in. of compacted gravel or crushed rock. Lap the sheet 6 in. at all joints and turn it up 6 in. at the walls. Do not use building paper; it is too porous to do a good job. The gravel and membrane will work together to prevent ground water from rising to and through the concrete.
- Cracking of concrete slabs can be controlled by:
  (1) placing the slab in small sections not over 20 feet in either direction;
  (2) placing every other section first; after hardening, place the sections in between (A typical placement sequence is shown);
  (3) installing a joint between the section that will transfer an edge load to the adjacent slab;
  (4) construction joints as shown; also expansion and control joints.
- Place a 4-in.-thick slab for normal use and 6-in.-thick slab where heavy wheel loads must be supported.

Wooden Floors

Wood can be used for floors that have to support light loads, such as in your poultry and swine housing and grain bins. A typical wooden floor uses plywood sheets placed over floor joists that span from outside wall to outside wall in a typical framework. If your floor is to remain dry and be long lasting, it is important that you place the joists and plywood off the ground, and that the area under the floor is open for good air circulation. Wooden floors are often made of exterior plywood panels ½ to ¾-in. thick, depending on the distance between joists and how heavy the loads on the floor will be.

Slotted Floors

Slotted floors are popular for livestock housing because they reduce cleaning time and effort. When your building has slotted floors, you will be able to easily remove dung from the pit below with liquid manure pump and wagon.

Slotted floors can be made of wood, concrete, steel, or aluminum. You should use steel or concrete slats in your swine or sheep housing; wood slats are fine for calf housing. Either wood or concrete will work well for your heifers and cows.

Expanded metal (steel) floors are becoming more popular, especially for sheep and swine housing. They permit free passage to dirt, waste, light, heat, and air. They provide sure footing, ample drainage and are inexpensive to install and maintain. For best results, investigate Wheeling's flattened mesh flooring for swine and Safe-T-Mesh for sheep.

Animal droppings pass more easily through slats with cambered tops and tapered sides such as wood or concrete slats. Steel or aluminum slats can be made from standard "T" or channel sections, although standard shapes are not often used because droppings will stick to them. Recommended slat spacing is shown.

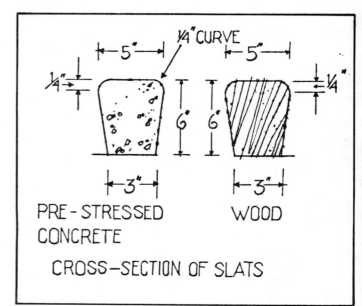

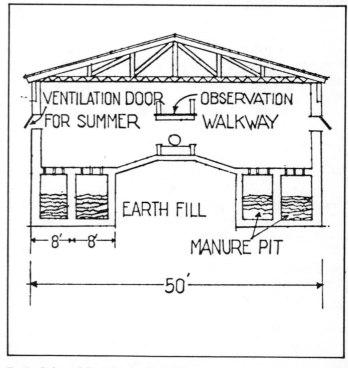

Typical slotted-floor layout for heifers.

Insulated Floors

You should consider insulating your floors for livestock housing only if warm floors are important, particularly in zones 1 and 2. Insulated floors reduce heat loss and provide a warmer surface. This is most important when animals are small and bedding is not used. Baby pigs often die shortly after birth if they are chilled on a cold concrete floor, and a common cause of mastitis in cows is the chill from cold floors. Also, frozen milk on your dairy house floor can be a serious safety hazard.

Insulation on your building's perimeter is important where animals are penned against the outside wall or when condensation on the floor next to the outside wall is objectionable.

There are two methods used for installing perimeter insulation: One method is to install a waterproof rigid foam insulation between the edge of the slab and the foundation and let it extend down along the foundation; the other method is to place the insulation under the floor instead of against the foundation.

The thickness and depth of insulation depends upon how low the winter temperatures in your area, but the top of the insulation board should always be placed level with the top of the slab.

Heated Floors

If you live in a very cold area, you should consider heated floors. Electric heat cables and hot water systems can be embedded in the concrete slab. Be sure that the concrete floor is properly insulated to reduce heat loss. Heated floors are most commonly used in farrowing pens.

Building Frames

Building frames support the building's loads and transfer them into the foundation and ground. The major components include:

(1) studs (or columns), which run from the foundation or ground up to the eave—they support the roof and siding and transfer loads to the foundations;
(2) girts, which run across the frame from side to side and connect the studs or columns—they give the frame lateral stability and the siding is attached to them;
(3) trusses, which are frameworks made up of several straight pieces that are usually joined together so that they form a series of triangles—trusses can be used for roof framing or for columns, and lightweight trusses used for roof framing are called joists;
(4) beams (or rafters), which are single horizontal or pitched pieces that resist vertical loads and support the roof;
(5) purlins, which are horizontal pieces that run across the beams, rafters, or roof trusses connecting them—they give added lateral stability and provide a connecting point for roofing materials.
(6) the sill, which is a piece of wood laid level on top of a masonry or concrete foundation wall—also called a splash plate, though if you use pole framing it is connected at ground level to the vertical poles;
(7) the plate, which is a horizontal piece at the top of the outside walls—the rafters, roof beams, or roof trusses are connected to it.

Types of Building Frames

There are four general types of frames used in farm buildings: platform, balloon, post, and rigid frames.

Platform frame. Composed of "independent" floors, a platform frame can support heavier loads than other framing methods and will settle uniformly if your lumber shrinks. In a multistory building each floor is supported by columns or studs only one story high.

Balloon frame. Also called a stud-and-out frame, it has continuous studs from foundation sill to eave plate, and all floors are supported from these continuous studs. Balloon frames are more resistant to wind damage than platform frames, but are less rigid and more easily consumed by fire.

Post frame. This frame uses heavy, pressure-treated pentachlorophenol square posts set in the ground 3 to 5 ft. deep. The advantage of post frames is that you do not need foundations. Sills are replaced by a horizontal railing (splash) board, which serves as a nailer for siding and is attached to the outside of the poles.

SECTION II: Setting Up the Basic Farm Structure

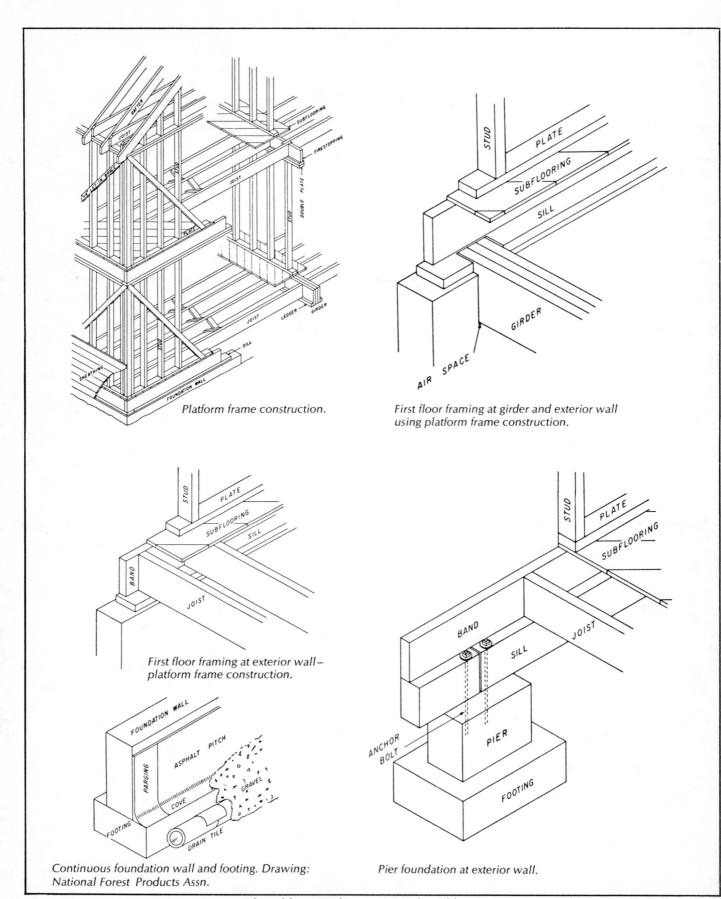

Platform frame construction.

First floor framing at girder and exterior wall using platform frame construction.

First floor framing at exterior wall— platform frame construction.

Continuous foundation wall and footing. Drawing: National Forest Products Assn.

Pier foundation at exterior wall.

Art reproduced from Finding & Fixing the Older Home, a Successful book by Joseph Schram.

Because square post frames are so simple, they can be installed faster, making them more economical than other framing methods. But they can only be used for lightly loaded buildings, unless footings are placed under the square posts, significantly increasing the cost.

Post-framed buildings are becoming more and more popular. They are inexpensive and they are particularly well suited for modern single-story buildings, such as machinery storages, sheep and beef cattle housing, loose housing for dairy cattle, separate hay storage and hay resting on the ground.

Rigid frames. These are normally used when you choose steel for the framing material. Girts and purlins can be steel channels or wood.

Rigid frames made from steel are more durable and stable than wood, and are almost fireproof. But they are heavier than wood frames, so you must build a larger foundation and you will need more labor and equipment to put them up. Steel frames are best for buildings with heavy loadings, such as grain storage buildings.

Wood Framing Details

Sills. Sills are usually made of 2-inch nominal thickness wood, and run anywhere from 4 to 12 in. wide. You install them continuously around the edge of the building on top of a masonry or concrete foundation. A single sill of 2-in.-thick lumber is commonly used. It is important to anchor the sills securely to the foundations because they tie the walls to the foundation so that the building will stand up to heavy winds. The most common anchoring method is to extend anchor bolts through the sill thickness into a concrete or block foundation. The bolts should penetrate the foundation at least 12 in. and should be spaced no more than 6 ft. apart. Use ⅝-in.-diameter bolts or larger for heavily loaded buildings, but for lighter, one-story buildings ½-in. bolts on 8-ft. centers are satisfactory.

With a concrete foundation you cast the bolts in the concrete as it is placed.

If you use a block foundation, bolts can be placed in the block core and mortared in place after the wall is up.

Studs. Studs (columns) must support floor and roof loads from above. They must be stable enough to withstand wind forces from the outside as well as pressure from materials stored inside. Studs are also the vertical parts of the framework to which siding materials are fastened.

Wall studs in most farm buildings can be 2-by-4 in. lumber installed on 24-in. centers. For two-story buildings, 2-by-6-in. wall studs on 24-in. centers may be necessary.

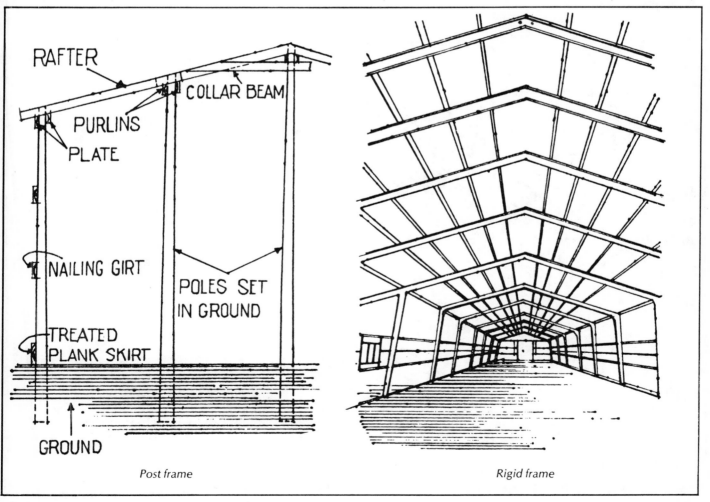

Post frame   Rigid frame

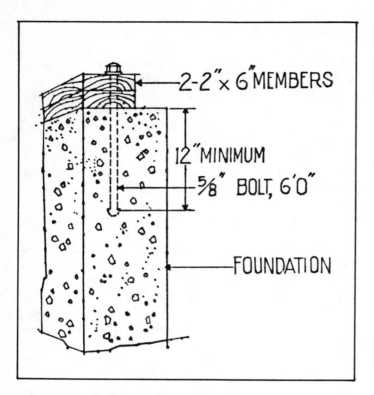

Bolts are cast in the concrete in foundations.

Corner studs are usually doubled and if a wall lining is placed inside, three studs are used to provide a place for nailing.

The bottoms of the studs must be securely anchored to the sill and the tops of the studs must be anchored to the plate. Toenailing was once the most common way to fasten studs to sills. But toenailing may not make the joint strong enough. A better method is to use "L" straps, angle "U" straps, gusset plates, or TECO fasteners. Any one of these fasteners will give you a strong, wind-resistant joint.

Plates. The plate is usually a 2-in.-thick piece of lumber the same width as the stud it rests on. It serves as a cap for the wall and a base for rafters, roof beams, or trusses. The plate is anchored to the studs by gussets, angle clips, strap iron, or TECO fasteners, like those used for fastening studs to sills.

In post framing, two 2-by-8-in. plates are nailed to each side of a pole. The 2-by-8 in. inside plate should be set higher to accommodate rafter slope (see p. 45).

Bracing. Wall frame bracing is necessary to prevent your building from "racking", or going out of plumb when it is under later pressures.

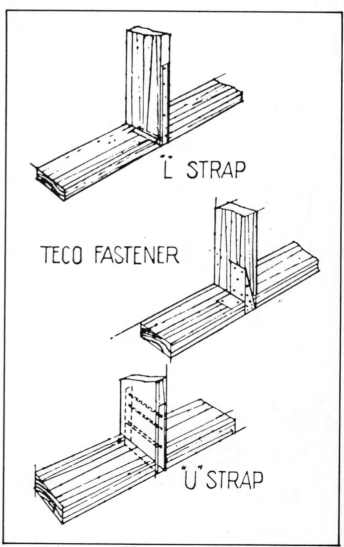

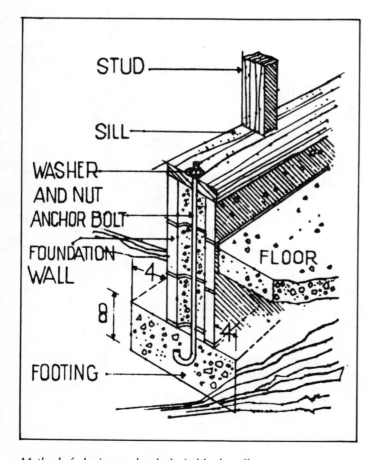

Method of placing anchor bolts in block wall.

Diagonal corner bracing is usually used to prevent racking. A 1x4 is placed in each building corner at a 45-degree angle to the studs. One end is firmly attached to the plate and the other end to the sill, and nailed to all intermediate studs. Bracing can be placed on the inside surface of the studs, or can be cut in or set in. Flat steel bands are also used in place of wood.

Doors and Windows. Door and window openings require special framing to support the building directly above the opening. They are usually provided by cutting one or more studs, which forces the adjacent studs to carry much more weight.

If only one stud is cut to make a door opening, the building will probably be adequately supported if you put double studs on either side of the opening, doubled 2x4 headers at the opening's top (and bottom, if a window opening), and short studs to the plate and sill of a window opening.

If more than one stud is cut, then the top and bottom headers and adjacent studs must be heavier and should be engineered specifically for your building.

Typical door and window frames are shown.

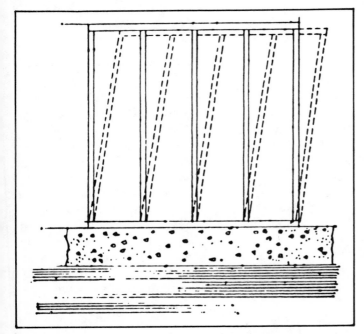

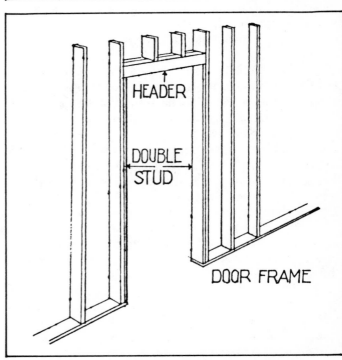

Unbraced wood frame has a tendency to rack.

43

Roofs

Roofs come in many types, differing mainly in their shape and slope. The roof type you select will depend on how you want your building and farmstead to look. Eight roof types are illustrated here. Advantages and disadvantages of each are discussed below.

Flat. Flat roofs are simple to build, but upkeep is high because water cannot run off them easily and they often spring leaks.

Shed. Shed roofs are used for leantos on existing buildings, single-story poultry houses, and open sheds for sheep, cattle, or swine. They are the easiest roofs to construct and maintain.

Gable. Gable roofs are probably the most popular type of farm roof. They are simple and easy to construct and maintain, and are used when roof spans are too large for shed roofs.

Hip. Hip roofs are more complicated to build than gable roofs and are therefore more expensive. They have no real advantage over gable roofs, except that gable end walls are eliminated. They are often used on garages and machinery storage buildings to give variety.

Monitor. Monitor and semimonitor roofs have lost their former popularity because they are complicated and expensive to build. But they do give additional height for storage, and added vertical wall space for ventilation and natural lighting.

Gambrel. Gambrel roofs offer more space for overhead hay and feed storage. For this reason, they are used on two-story dairy barns and other livestock shelters where overhead storage is desirable. But because special and complicated truss framing is needed, gambrel roofs are expensive. Their popularity is also dropping because more and more people favor one-story buildings and ground-level feed storage.

Arched. Arched roofs are shaped just like gambrel roofs, but they are made of prefabricated laminated rafters to simplify construction. They have most of the same advantages and disadvantages as gambrel roofs.

Roof Rafters. Rafters and roof joists (trusses) support the roofing material. In shed roofs, rafters connect plates on the low and high walls. Gable roof rafters connect plates on outside walls to the ridge. Hip roof rafters connect outside wall plate to ridge beams and to hip rafters.

Gable and hip roof rafters are normally notched into a "seat-cut" at the plate connection and a "plumb-cut" at the ridge connection to assure maximum contact at each joint. Seat-cuts are not necessary if purlins are installed between the rafters.

Roof trusses are used for roof spans over 24 ft. when columns inside the building are undesirable; thus, 24 ft. can be clear-spanned using gable or hip roofs with standard rafters or rafters and purlins. See the Warren (a) Fink (b), and Howe (c) truss designs for clear spans up to 36 ft. on page 47.

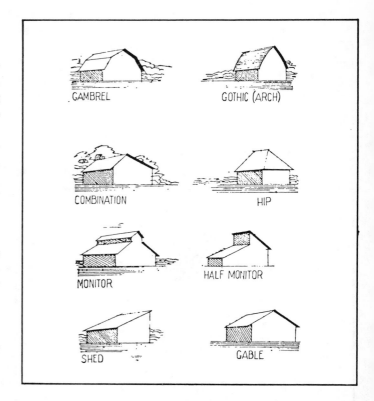

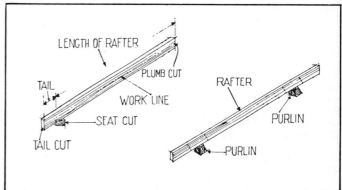

Connecting the Wood Members

Joints connecting wood members should be as strong as the members themselves. It is poor economy to use heavy timbers and then connect them so lightly that the joints are the weakest part of your building. Remember, your building frame is no stronger than its weakest link.

Common nails are by far the most usual connectors, and are adequate for most frame joints. Special nails, such as barbed, spiral, grooved, square, screw, and triangular-shanked, have greater holding power than common nails, but are more expensive and normally unnecessary.

Common nails are sized by diameter, length, and number per pound; sizes are designated by penny, symbolized "d".

You should drive your nails into wood at right angles to the grain. Driven parallel to the grain, the nail will lose at least 50 percent of its holding power.

Building Basics

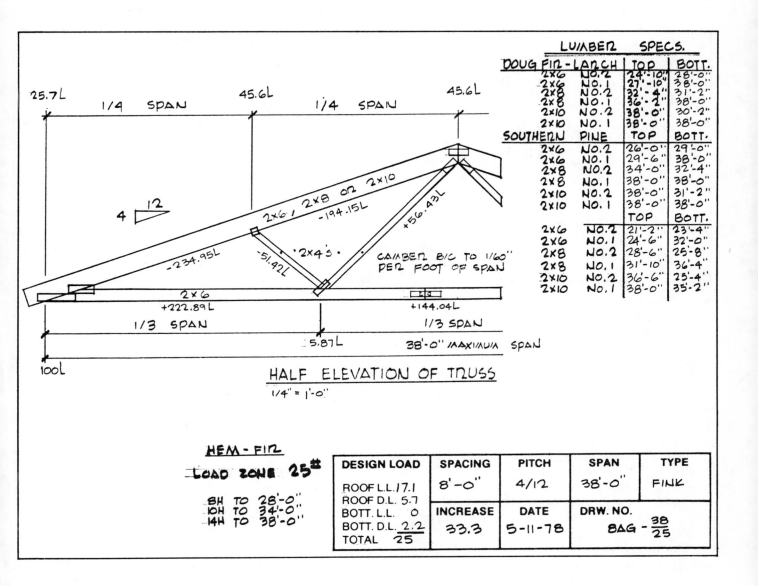

To estimate the nail size you need for each joint, figure that the nail must penetrate into the second receiving piece of wood for at least two-thirds of a softwood's thickness and one-half of a hardwood's thickness.

As an example, suppose you want to nail two softwood two-by-fours together, long side to long side. Each piece of wood is 1 5/8 in. thick for a total thickness of 3 1/4 in. A nail should go all the way through the first piece of wood, 1 5/8 in., and two-thirds of the way through the second piece, 1 1/8 in. When you add the two together you get 2 3/4 in., which is the nail length you need. Note that an eight-penny nail is 2 1/2 in. long and that a ten-penny nail is 3 in., so use ten-penny nails to be on the safe side.

Split-ring connectors are used for heavy-duty joints, such as at rafter and truss peaks, and for connecting rafters to poles in pole-framed buildings. The rings are placed in pre-cut grooves on the faces of the wood to be connected. The two pieces of wood are then bolted.

Split rings are available in two standard sizes: 2 1/2 in. with a 1/2-in. bolt for use with 2-in. lumber; and 4 in. with a 3/4-in. bolt for heavier lumber. Grooves are cut in the wood with a grooving tool attached to an electric drill.

The tooth-ring connector is used for lighter framing than the split-ring connector. You don't need power tools to do the grooving, because the toothed ring is driven into the wood as the bolt is tightened.

To make a toothed-ring connection, bore the bolt hole first and then press the ring onto the wood's contact faces. As the bolt is tightened the ring will penetrate the wood.

Selecting Lumber for Your Building

The higher the density, the greater the lumber's strength

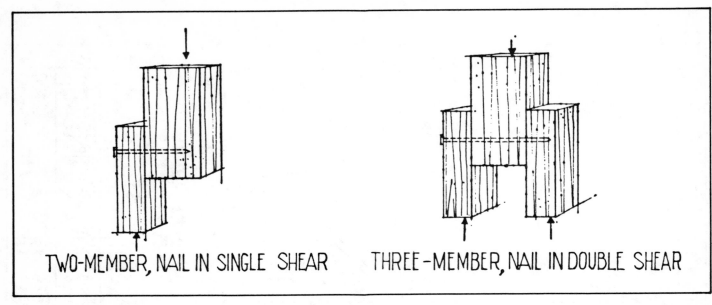

TWO-MEMBER, NAIL IN SINGLE SHEAR   THREE-MEMBER, NAIL IN DOUBLE SHEAR

to withstand bending and compression. Density is used to measure grade and ultimate use. The denser woods are heavy-load-bearing girders and posts; the lighter woods serve for light framing and studs. Where heavy loads are expected, use a wood with at least six annual rings to the radial inch.

Defects, such as large or loose knots, excess sapwood, checks, or decay, will reduce the lumber's strength, especially where the wood is in tension, for example, at the side of beams.

Dry or seasoned wood is much stronger and stiffer than green wood. It will not shrink after it is in place. If you use unseasoned wood for a building frame, you run the risk of having the frame shrink and twist, check and warp. This will cause the roof and walls to leak and, more important, you may end up with a basically unstable building.

Structural wood, used for building frames, is usually air dried until it contains only 12 to 20 percent moisture. Kiln-dried wood has only 6 to 12 percent moisture; it will shrink less, which is important to reduce or eliminate frame racking.

Economy is always important, and you can save money by selecting the right kind and grade of wood. Remember that denser, more perfect grades are stronger but more expensive than light, soft woods. They are also harder to work. In general, use the dense woods in beams, girders, rafters, and joists. Use lighter woods for studs and other lightly loaded members.

Lumber grades are standardized throughout the United States; grades can be categorized as yard, structural, factory, and shop lumber.

Yard lumber includes boards and siding, planks and joists, and ordinary 2x4-, 3x4-, 4x4-, 2x6-, and 3x6-in. studs. Yard lumber is graded according to presence of knots, sap, and other blemishes, into six grades: A, B, and C; and Number 1, Number 2, and Number 3 common.

Structural lumber is heavier than yard lumber and is graded according to its density, strength, and stiffness. It is used for beams, girders, posts, and sills over 5 in. thick (least dimension). It is graded by five general classifications:

(1) dense select—Douglas fir and southern pine;
(2) select—Douglas fir;
(3) select—other softwood species except southern pine;
(4) dense common—Douglas fir and southern pine;
(5) common—all softwood species.

Factory- and shop lumber is very high quality, blemish-free wood used for making such items as furniture. It is almost never used in building frames.

In general, you should follow these three rules when you select your framing and structural lumber:

(1) for lumber 1½ in. thick (and less) and for all studding, use Number 1 common yard lumber;
(2) for joists and rafters, use common structural;
(3) for girders, heavy beams, and posts, use structural or prime structural.

How to Make Your Wood-Framed Building Stiffer and Stronger

Beyond this basic "how to" data for building wood frames, the following tips can help to make your frame stronger and longer lasting.

- Ten-penny nails instead of eights will increase the stiffness of horizontal sheathed walls 50 percent and their strengths up to 40 percent. (They do not, however, improve diagonally sheathed walls.)
- Putting three or four nails in horizontal sheathing instead of two does not improve the wall, but will add

30 percent to 100 percent to the stiffness of diagonally sheathed walls.
- Side- and end-matched sheathing not butting over the studs is as stiff and strong as the sheathing that butts over the studs.
- Green lumber is half as stiff and 70 percent as strong as seasoned lumber.
- Herringbone bracing (or bridge bracing) between studs has little value, but diagonal wall braces cut in or let in between studs add 60 percent to stiffness and 40 percent to strength.

General Notes and Comments On Pole Structures

This section will give you a perspective on the layout and construction of pole-type structures. At the present time there is probably no lower cost structure that can be erected in the construction field. Some of the advantages of pole structures are:

(1) low initial investment;
(2) can be used as temporary or permanent shelter;
(3) can be erected rapidly in almost any reasonable weather conditions;
(4) requires a lesser degree of skilled labor for satisfactory erection;
(5) when properly constructed, it is basically a low-maintenance, easily maintained structure;
(6) it is easily expanded and adapted for future needs.

What is a Pole Building?

A pole structure is a building that utilizes treated wood members for the foundation, vertical support members and base skirt boards rather than concrete, masonry, or metal framework. Instead of extensive excavation and poured concrete, holes are drilled and poles set in place much like telephone poles. Where soil conditions require, an 8-in.-thick concrete pad is sometimes poured in the bottom of each hole before setting poles.

Life Expectancy

Pressure-treated with proven preservatives, the poles or sawn timbers used for the frame of the structure can give long life expectancy. Properly treated poles have proven their worth by the service that utility companies all over the country have come to expect from them. Latest estimates indicate that 45 or 50 years is not unreasonable for the service life of a pressure-treated wood member. The treated lumber that is on the inside of the building has added protection and will last longer.

Job Preparation

Listed below are some of the time and cost-saving areas that result in the exceptional low cost of a pole structure.

Grading	Required only to prepare level spot for building floor itself
Excavation	Drilling or digging of pole holes
Foundation Work	None required
Pre-Framing	None normally required
On-the-job engineering	None required
Erection Equipment	Usually only light equipment needed

To further reduce on-site labor and framing, prefabricated wood trusses have been used more and more in the erection of pole type structures. It is not uncommon for a pole structure with wood trusses to be utilized for structures with a 40-ft., 60-ft. and even 80-ft. clear span width. The length possibilities of the building would be unlimited. Some builders prefer to fabricate their own trusses, using casein or resorcinol resin glue and nails, while others prefer to use the prefabricated and pre-engineered trusses available from their materials supplier. We will show briefly some of the data necessary in planning, layout and building of trusses merely to add perspective to your overall planning of a project. It is suggested that you coordinate final plans with your building materials supplier and your Agricultural Engineering Extension Service of your state university. Your County Agent can be of help in this area.

Embedment Depth

Post Size	Poor Soil	Average Soil	Good Soil
4 × 4	4'—4"	3'—6"	2'—8"
6 × 6 or 4 × 6	5'—6"	4'—6"	3'—6"
8 × 8 or 6 × 8	7'—0"	5'—6"	4'—6"

Pole Dia. (Top)	8'0" Eave			14'0" Eave			20'0" Eave		
	Poor	Average	Good	Poor	Average	Good	Poor	Average	Good
4"	5'5"	4'4"	3'5"	5'11"	4'8"	3'8"	6'4"	5'0"	3'11"
5"	6'4"	5'0"	3'11"	6'7"	5'3"	4'2"	6'11"	5'6"	4'4"
6"	6'11"	5'6"	4'4"	7'3"	5'9"	4'7"	7'7"	6'0"	4'9"
7"	7'7"	6'0"	4'9"	7'11"	6'3"	5'0"	8'3"	6'6"	5'2"
8"	8'3"	6'6"	5'2"	8'6"	6'9"	5'4"	8'10"	7'0"	5'7"

SECTION II: Setting Up the Basic Farm Structure

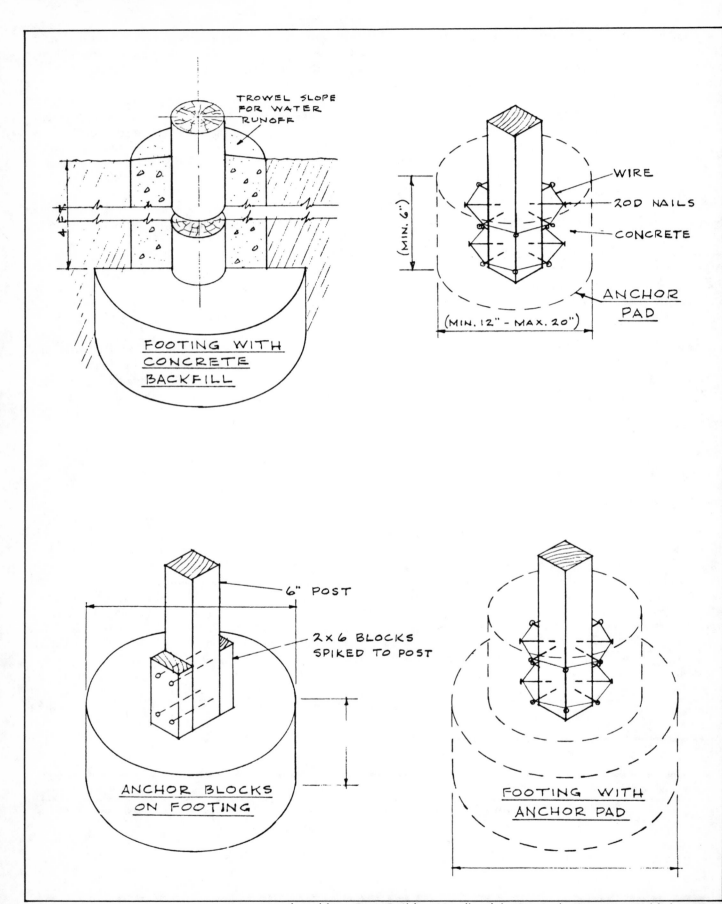

Art reproduced from Farm Builders Handbook *by R.L. Lytle (Structures Publishing Co.).*

The plans, drawings and specifications shown were basically assembled from standard date and pole barn plans available from the Pennsylvania State University and the United States Department of Agriculture. Many of these plans are adaptable throughout many portions of the United States, but in areas where greater snow loads or more severe weather conditions prevail, local guidance and engineering should be obtained. The basic principles of pole structures are the same wherever they are used. It is only certain spacings and structural details that may require modifications. For the purpose of calculations, all roof loads on this plan have been engineered and designed for a 30-pound-per-square-foot combined load.

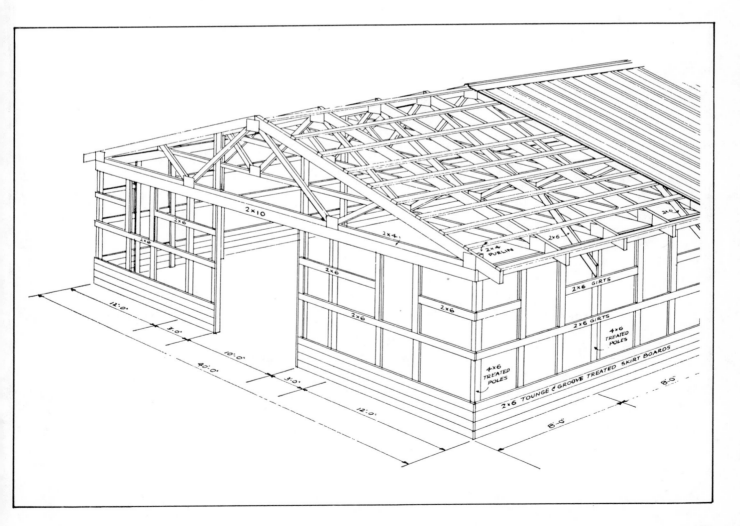

SECTION II: Setting Up the Basic Farm Structure

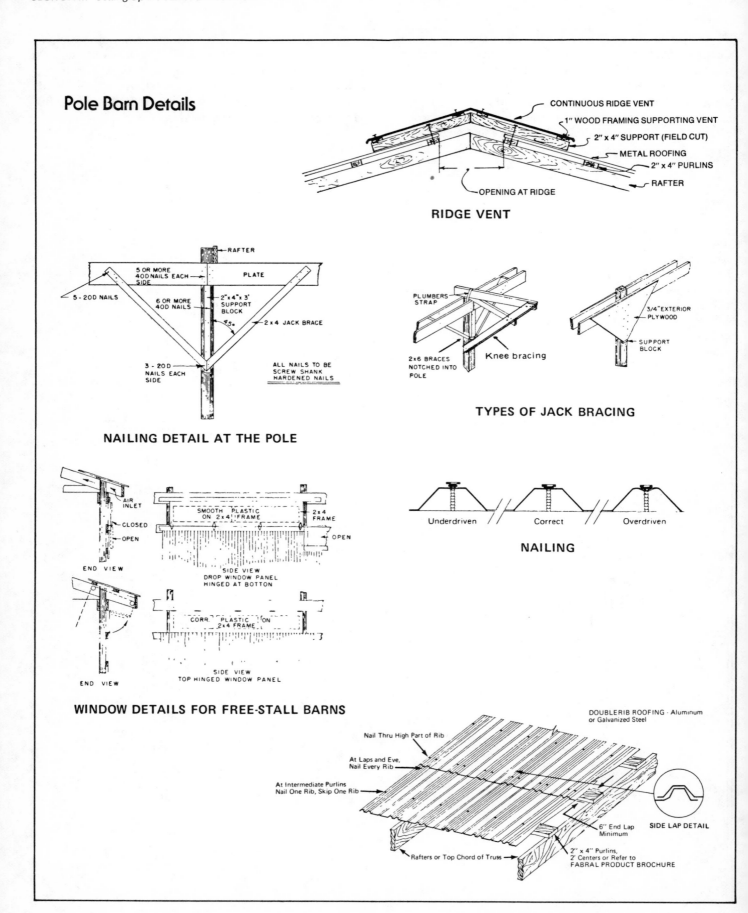

6. Environmental Controls

Studies show that it is important to control the temperature, humidity, and ventilation in livestock buildings because extreme high and low temperatures affect production and health. You can regulate temperature, humidity, and ventilation rates with simple and readily available mechanical equipment. This mechanical equipment, in turn, is controlled by sensing devices such as thermostats and humidistats.

Whether you need this equipment or not depends largely on how it will affect production and profits. Therefore, you should carefully consider your needs before spending the substantial amount of money required for environmental control. And we recommend that you base your needs on an understanding of how your farm animals will react to specific environments and how these environments can benefit you.

How Farm Animals React to Controlled Environments

Your farm animals will react to different environments in different ways.

Dairy Cows

Dairy cows' milk production will not seriously increase or decrease between 0 and 80° F., provided temperature changes are gradual. When the temperature goes above 80° F., however, your cows' milk production will go down rapidly as more moisture is thrown off by the animal to cool its body.

Studies show that if you maintain an even 50° F. in your dairy barn, milk production will be at its highest possible level. However, you must be practical, so we recommend that your dairy barn temperatures in zone 1 range between 35° F. and 45° F., in zone 2 between 45° and 50° F., and in zone 3 between 45° and 55° F.

Agricultural engineers don't know for sure how humidity in dairy barns will affect milk production. Their recommendations, based on the experience of many farmers, are to maintain 75 percent to 80 percent relative humidity to get the most production.

Ventilation requirements in your dairy barn range from 100 cu. ft. of air per minute for each 1,000-pound animal when the weather is hot and humid, to 40 cu. ft. per minute per animal in cold and dry weather.

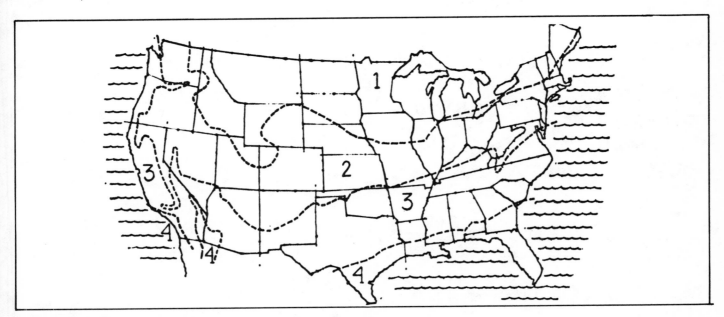

Farm building-zone map, based on January temperatures and relative humidity.

SECTION II: Setting Up the Basic Farm Structure

Laying Hens

Laying hens will produce more eggs when the temperature in your laying house is 50° to 55° F. But egg production is not affected too much unless the temperature drops below 45° F. or rises above 65° F. However, sudden and severe temperature changes will make your egg production drop significantly.

The humidity in your laying house should be kept between 75 and 85 percent. If the humidity and temperature get very high, expect your hens to produce far fewer eggs.

To ventilate your laying house, plan on moving 4 cu. ft. per minute for each of your hens when the temperature is 85° F. or higher. Use adjustable ventilation methods so that you can reduce the flow during winter months.

Beef Cattle

Beef cattle are hardy and do not normally need controlled environments. They can stand very low or high temperatures and humidity without losing weight or slowing down their rate of weight gain. Ventilation requirements are only 50 to 60 cu. ft. of air per minute per animal, a low rate that is easily satisfied by simply opening the barn door.

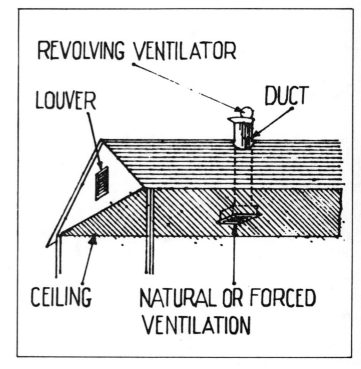

Sheep

Sheep can also tolerate low temperatures. They do not need any heat except when they are lambing, and even then a light bulb suspended over the enclosed claiming pen will be enough. Humidity control is not necessary.

Your sheep will need some draft-free ventilation at lambing time and after shearing.

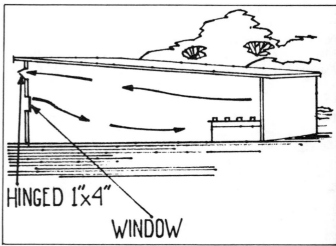

Swine

Swine from 70 to 144 pounds will grow most economically at 75° F.; from 166 to 260 pounds they'll grow best at 60° F. They do not need humidity control.

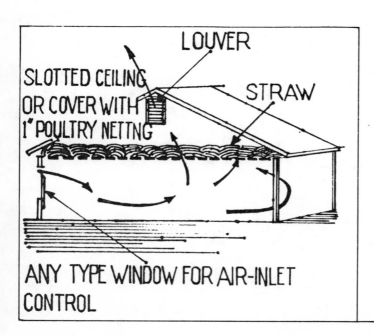

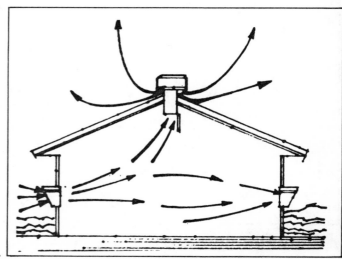

Ventilating Methods

Two methods for ventilating your buildings are commonly used: gravity and forced fan.

Gravity ventilation works when cold outdoor air forces the warmer indoor air up and out of the building. Because gravity ventilation does not use any mechanical equipment, it costs almost nothing to operate. However, it does not enable you to control air volume and temperature very exactly. Several schemes to install draft-free gravity ventilation are shown below.

Forced-fan ventilation can be either blow-in (pressure) or blow-out types. If you install forced-fan ventilation instead of gravity ventilation, you will get more positive air movement and temperature control, but it will cost more to install and to operate. It is best to rely on recommendations of ventilation equipment dealers or specialists for this type of system.

Blow-in and blow-out forced-fan ventilation types can often be combined by installing fans that can be reversed. This will make your ventilation system more useful to you because it can then be used in all seasons. The fan on a side-wall that draws air from the attic in winter (blow-out) can be reversed in summer to bring in air from eave openings (blow-in) and return it out through louvers.

In a house without a ceiling, the summer air is drawn through side-wall fans and exhausted through adjustable ridge and side-wall vents (blow-in). In winter the fan is reversed and air is brought in through the ridge vents and exhausted along the side-walls (blow-out).

Temperature Control

In most of your farm buildings and in most climates, heat from the sun and your animals is enough to maintain the most desirable winter temperatures, and a well-designed ventilation system can maintain satisfactory summer temperatures. However, in zones 1 and 4 where you can expect extremely low or high temperatures, you might need some type of cooling or heating help.

Your heating and cooling needs will depend upon your climate, the temperature you need in the building, amount of insulation, and ventilation rate.

As a general rule-of-thumb, however, you can calculate your approximate heating or cooling needs by figuring your building's volume in cubic feet, and then multiplying the total cubic feet times .018 BTU (British Thermal Unit) per degree of temperature rise or fall.

Let's take an example. Your building is 100 ft. by 50 ft. by 20 ft. high, or 100,000 cu. ft. You want to raise the temperature inside the building 40° F. higher than the lowest outside temperature you can expect (or lower it 40° F.). To figure the approximate heating (or cooling) capacity needed, multiply:

100,000 cu. ft. × .018 × 40° F. = 72,000 BTU per hr.

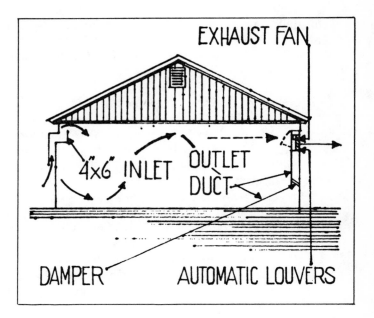

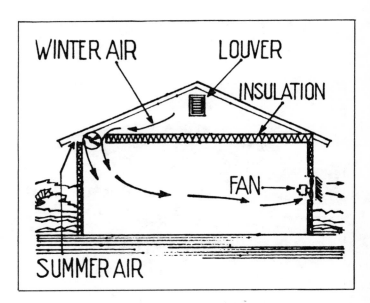

Your particular situation probably will raise or lower this figure, so it's best to check with a local heating and air-conditioning contractor who has experience with your area and with similar buildings.

The most common heating units are gas- or oil-fired heaters with fan-forced air circulation, mounted on the floor or suspended from roof trusses.

Cooling units are usually standard electric- or gas-powered air-conditioners or evaporative coolers. Fog nozzles can be used where humidity does not go over 65 to 60 percent. Blocks of ice placed in front of blow-in ventilation fans are often used but this method is hard to control and not very effective at best.

A newer concept called solar orientation places your building to take full advantage of solar radiation for heating the building in winter and for lowering the sun's heat entering the building in summer. If you understand and use solar orientation, it might be possible to eliminate or reduce the need for auxiliary heating or cooling. You should cosider these four important factors.

- Face widows south to let your building absorb as much heat as possible from the winter sun.
- Make your windows larger than normal so that more winter solar energy can enter the building.
- Use insulating (double-pane) glass in all your windows to reduce winter heat loss. Most rays from the sun have relatively short wavelengths. They are easily transmitted through insulating glass, absorbed by solid objects in the building, and changed over to heat. Insulating glass also will reduce loss of this heat.
- Install sun control devices whenever possible for summer use. Build extra-large roof overhangs, and use visors or shades to lower window and wall exposure to the summer sun.

Insulation

Controlling the environment in your building is often easier and more economical if your building is insulated. Insulation, if it's properly applied, will reduce heat loss or gain through walls and roofs and eliminate condensation on inside wall surfaces.

You can choose from many types of available commercial insulation materials, such as loose fill, batts and blankets, insulation boards, plastic foams, and reflective foil. All materials are rated by their corresponding "R" value. An "R" value of 4 is twice as good as 2.

The value is sometimes listed per inch of thickness. For example, 1 in. of glass fiber or mineral wool has an "R" value of about 3.7 and 1 in. of fiberboard has only about 2.6.

Sometimes the "R" value is listed for a material's full thickness, such as 7.4 for 2 in. of glass fiber, or 5.2 for 2 in. of fiberboard.

Regardless of which insulating material you choose, be sure you put a vapor barrier on the warm side of all insulated walls and roofs. This will prevent water vapor from condensing in the insulation, wetting it, and destroying insulating value. Polyethylene film, aluminum foil, or several coats of aluminum-flake, applied to the insulation's inner unbroken surface are all vapor barriers.

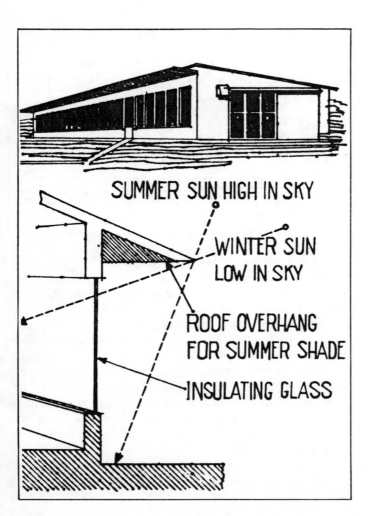

7. All About Roofing Farm Buildings

Many types of roofing are available for farm buildings: asphalt or asbestos, cement shingles, roll roofing, galvanized steel, aluminum, wood shingles, slate and others. They vary in durability, fire resistance, insulating value, and other properties. Make your selection carefully whether you are covering a new building or reroofing an old one.

For maximum service and protection, roofing must be installed properly and kept in good repair. Improper installation and poor maintenance can result in leaks or other troubles that will shorten the life of the roofing.

Roofing materials are commonly sold by the square (100 sq. ft.). The number of squares needed will be determined by the area of the roof in square feet.

Shown are four types of common roofs used on farms and a simplified method of determining the area of each. The roofs shown are plain. If your roof has dormers, chimneys, and valleys, determine the area by sections and add.

Extra material is required for overhang at eaves and gables and for fitting around chimneys, dormers, and valleys. Include this in your estimate. Allow also for waste.

Selection

The roof covering for a building should be carefully selected. Some important considerations are roof slope, weight of roofing material, cost, fire resistance, appearance, and location.

Roof Slope

Shown is the minimum roof slope on which various types of roofing should be laid using the standard end or side lap. If the slope is less than that indicated there will be danger of leaks.

Weight

Roofing materials vary in weight. The table shows the approximate weight per square of different types. If the roofing is too heavy for the framing, sagging may occur. A roof that sags is unsightly and hard to keep repaired.

Cost

Roofing materials vary widely in price. Cost in roofing however, involves more than the cost of the materials. Labor, decking, scaffolding, and other factors make up a large part of the cost.

In selecting roofing from the standpoint of cost, keep in mind that good quality, long-lived roofing should be used on the permanent buildings, even though the first cost is high. Once maintenance, repair, and replacement are considered, low-quality roofing can be more expensive in the long run. Long-lived roofing is also warranted when the cost of applying the roofing is high in comparison with the cost of the materials, or when access to the roof is hazardous.

If you are near the supply center, good-quality roofing may be available at lower-than-normal cost. For example,

Lightweight felts—average size and weight of commercial packages

Material	Width	Area per roll	Weight per roll
	Inches	Square feet	Pounds
15-pound asphalt or tarred felt..	32 and 36	432	60
30-pound asphalt or tarred felt..	32 and 36	216	60
Slater's felt................	36	500	32
Sheathing felt..............	36	500	35
Red rosin-sized sheathing paper..	36	500	20, 25, 30, and 40

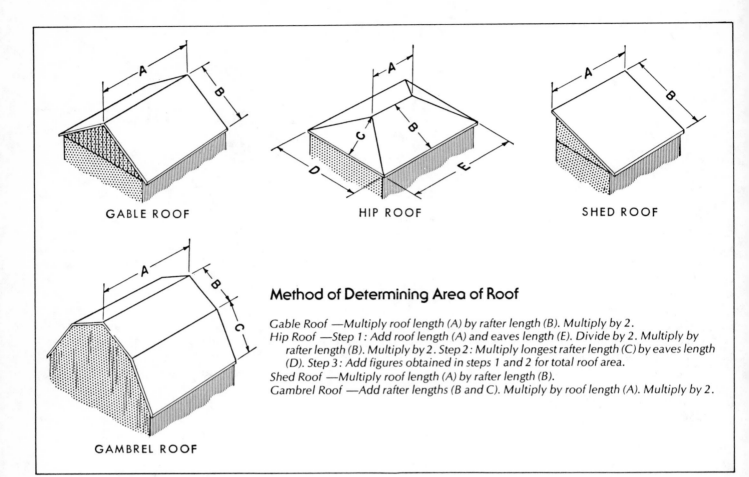

Method of Determining Area of Roof

Gable Roof —Multiply roof length (A) by rafter length (B). Multiply by 2.
Hip Roof —Step 1: Add roof length (A) and eaves length (E). Divide by 2. Multiply by rafter length (B). Multiply by 2. Step 2: Multiply longest rafter length (C) by eaves length (D). Step 3: Add figures obtained in steps 1 and 2 for total roof area.
Shed Roof —Multiply roof length (A) by rafter length (B).
Gambrel Roof —Add rafter lengths (B and C). Multiply by roof length (A). Multiply by 2.

slate is one of the most durable roofing materials and usually is one of the more expensive. However, near a quarry the price may be comparable with that of lower-quality less-durable roofing.

Fire Resistance

Roofing materials vary in fire resistance. Slate, asbestos-cement shingles, and metal roofings are the most fire resistant. Others, such as asphalt shingles and roll roofing, provide satisfactory protection, if they are of good quality and are kept in good condition. A Class C label indicates shingles were UL-tested for light fire exposure, and they will not readily ignite or support the spread of fire. A Class A UL-label is even more fire-resistant.

The dwelling and other important buildings should have a fire-resistant roof covering, if possible. Buildings closely grouped together—less than 150 ft. apart—also should have fire-resistant coverings. If one catches fire, the danger of fire from flying sparks to the other buildings will be minimized.

Appearance

The roofs of buildings in a farm group should harmonize in color even though they may differ in coutour or design.

Light-colored asphalt roofs usually absorb less heat than darker-colored roofs.

Location

Along seacoasts the air is saturated with salt; around industrial works it may be polluted with fumes. The salt and fumes can corrode galvanized or aluminum roofings and shorten their life. Steel roofing, even though galvanized, is particularly susceptible to such corrosion. If used, it must be kept well painted.

Estimating Roofing Needs

Having selected the type of roofing, you will need to determine the amount of roofing material needed. This should be done by someone experienced in estimating roofing. The above is offered as a guide; add 10% for dormers, valleys, overhangs, and waste.

Various kinds of rigid shingles are available. Wood, slate, and asbestos cement are discussed here, because they are most commonly on farm buildings. Others include clay shingles and tiles, molded-asbestos tile, and molded aluminum shingles.

Wood

Wood shingles, if of a durable species and properly laid, make a satisfactory, attractive, and well-insulated roof.

Different grades of shingles are on the market. The best ones are edge grained and all heartwood No. 1 grade. Shingles in all grades below No. 1 are flat grained or contain varying amounts of sapwood.

No. 1 grade southern cypress, redwood, and cedar shingles are the most decay resistant. No. 1 grade shingles are recommended for permanent roofs—especially for dwelling roofs. The lower grades of shingles are not economical for permanent construction but are suitable for temporary roofs and for sidewalls.

Wood shingles are made in lengths of 16, 18, and 24 in. Ordinarily they come in random widths of 2½ to 14 in. You can get shingles of uniform width—5 or 6 in.—but they are generally used for decorative effects on roof and sidewalls.

Deck. In warm, humid climates, wood shingles are commonly nailed to slats creating a roof that permits ventilation of the underside. A slat roof is light in weight and low in cost. The slats may be 1-by-4 in. strips spaced center to center, a distance equal to the length of shingle exposed to weather.

In cold climates, the shingles are usually laid over tight sheathing covered with rosin-sized paper. Slats with insulation board under them are sometimes used.

Method. One-fourth pitch is the minimum slope recommended for wood shingles. If the slope is much less than one-fourth pitch, it will be hard to keep the roof watertight.

On roofs of one-fourth pitch or steeper, lay the shingles as follows to provide a three-ply roof:

Shingle length (inches)	Length exposed to weather (inches)
16	5
18	5½
24	7½

On roofs of less than one-fourth pitch, lay the shingles as follows to provide a four-ply roof:

Shingle length (inches)	Length exposed to weather (inches)
16	3¾
18	4¼
24	5¾

Low-grade flat-grained shingles should be laid with the "bark" side exposed (the side that was nearest the bark in the tree). They will weather better and be less likely to turn up at the butt or to become waterlogged.

Split the shingles that are over 8 in. in width; atmospheric changes can crack wide shingles.

Double the shingles at all eaves and extend them about an inch beyond the edge. Space dry shingles ¼ in. apart, and green or wet ones ⅛ in. to allow for swelling in damp weather.

Fasten each shingle with two nails, one on each side, 1 to 2 in. above the butt line of the next course and not more than ¾ in. from the edge. Never nail in the middle—the shingle may split. Use three-penny rust-resistant nails for 16- and 18-in. shingles and four-penny nails for 24-in. shingles. Joints should be broken at least 1½ in. and all nails should be covered.

Check the coursing as the work progresses. The shingle

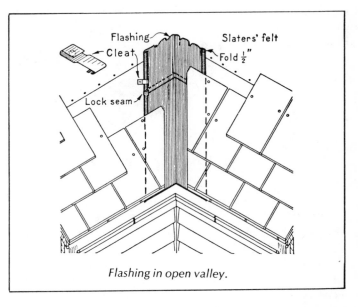

Flashing in open valley.

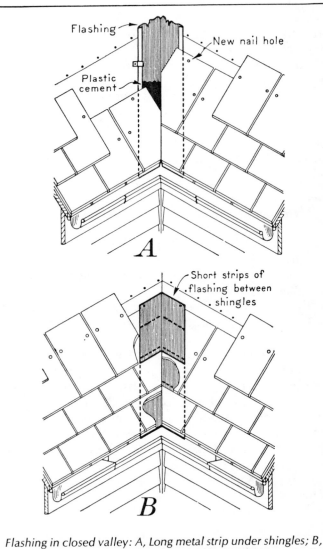

Flashing in closed valley: A, Long metal strip under shingles; B, short pieces of metal intermembered with shingles.

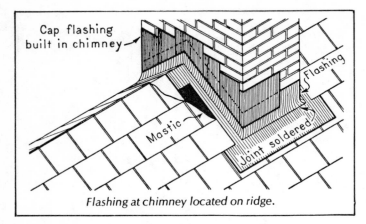

Flashing at chimney located on ridge.

rows must be kept parallel to the eaves to avoid uneven exposure of the last few courses.

Flashing. Painted sheet iron or "tin" is frequently used for flashing with wood shingles. However, more durable material, such as 26- or 24-gage galvanized metal of the highest quality or heavily coated IX flashing tin, is recommended.

Staining. Stains rich in coal-tar creosote have much more preservative value than those containing little or no coal-tar creosote. However, shingles treated with such stains cannot be satisfactorily painted; the creosote will bleed through paint even after several years' exposure. Shingles treated with stains containing little or no coal-tar creosote can be painted after short exposure to the weather.

Dipping is the best method of staining a shingle. Dip the shingle to within 3 in. of the tapered end. Brush coats may be applied for additional protection after several years' exposure.

Slate

Slate shingles make an attractive, durable, and fire-resistant roof covering.

They are available in different grades and in various colors. The best slates have a metallic appearance, do not absorb water, and are very strong. B-grade slate has sufficient durability for farm buildings.

Commercial slates range in size from 6 by 10 to 14 by 24 in. The more commonly used sizes are 8 by 16, 9 by 12, and 9 by 18 in.

Some dark slates fade to a lighter gray on exposure. This change in color is not always uniform and the roof may become unattractive. Certain green slates may become buff or brown after a few months' exposure. This change is sometimes considered desirable and it has no effect on the quality of the slate.

Slates are very heavy roofing material—700 to 900 pounds per square—and require strong roof framing. (A square of roofing is the quantity necessary to cover 100 sq. ft. of roof surface.)

Slate roofs are commonly installed by roofing contractors.

Asbestos Cement

Asbestos-cement shingles are made of asbestos fiber and

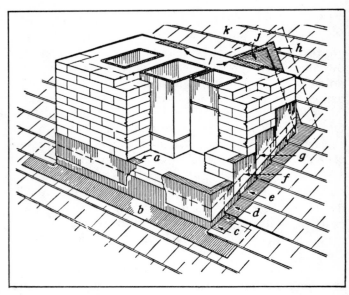

Flashing at chimney located on slope. Sheet metal (h), over the cricket, extends under the shingles (k) at least 4 inches and is counterflashed at l in joint. Base flashings (b, c, d, and e) and cap flashings (a, f, and g) lap over the base flashings to provide watertight construction. Provide a full bed of mortar where cap flashing is inserted in joints.

portland cement. They are strong, durable, and fire resistant.

They are available in a wide variety of colors and surface textures (including their natural color, which is similar to that of portland cement).

There is no standard by which to judge the merits of the many variations, and if you plan to use these shingles, select a type that has given good service in your locality.

Asbestos-cement shingles may be classed according to shape and method of application as follows.

AMERICAN-METHOD individual shingles are 8 in. wide and 16 in. long. They are laid like wood shingles and weigh about 350 pounds per square.

AMERICAN MULTIPLE shingles usually come in strips 24 to 30 in. long and 12 to 15 in. high. They weigh about 300 pounds per square. When laid, they give the appearance of smaller individual shingles.

SIDE-LAP, or Dutch-lap, shingles are approximately 16 by 16 in. and weight 265 to 290 pounds per square. They are laid with one-third or one-fourth side and top lap. One-third lap makes a tighter and more attractive roof. One-fourth lap makes a lighter-weight and lower-cost roof.

AMERICAN RANCH shingles are 24 by 12 in. and weigh 250 to 260 pounds per square. They are usually laid with one-sixth side lap (20 in. exposed) and one-fourth top lap (9 in. exposed).

Detailed instructions for laying asbestos-cement shingles are provided by manufacturers or dealers.

Flexible Roofing Materials

Asbestos-Cement Sheets

Corrugated asbestos-cement sheets are available for

covering roofs consisting of sheathing strips or purlins on top of rafters.

Dimensions of the sheets are: Width, 42 in.; length, 3 to 11 ft.; thickness, ⅜ in. or ¼ in. The sheets are corrugated at 4.2-in. intervals.

Following are installation details.

Deck. Sheets ⅜ in. thick may be laid on purlins spaced 45 ot 54 in. on center. Sheets ¼ in. thick may be laid on purlins spaced 30 to 42 in. on center. The exact spacing of the purlins required in each case will depend on expected snow loads.

Method. The sheets should be laid with a side lap of 1 in. corrugation and a minimum end lap of 6 in. Trim the corners of the sheets as shown to permit continuous lap.

Fasten the sheets to wood purlins with 3-in. ring-shank nails. Fasten them to metal purlins with special fasteners available commercially.

Apply asphalt mastic on each side of the ridge. Set a gasket material on the mastic. Then cover the ridge with a semicircular ridge cap.

Bituminous Roofing

Bituminous roofing, in one of its many forms, can be found on many farm buildings.

Bituminous roofing materials divide into three general classes: lightweight felts, roll roofing, and asphalt shingles.

There are, however, four types of bituminous roof covering—the three listed above plus builtup roofing.

Bituminous roofing materials have a felt base. They are made of rag felt or asbestos felt. Asbestos-felt roofing is more fire resistant than the rag-felt roofing.

Lightweight Felts. Lightweight asphalt-saturated or tar-saturated felts are used (1) under shingles or other roofing materials, (2) for builtup roofing, and (3) to cover low-cost buildings such as sheds. They serve only as a very temporary roof covering, however, as they are easily torn by the wind.

The table shows the several kinds of felts available and the average size and weight of commercial packages. Rosin-sized paper also is listed in the table. However, it is not felt material; it is merely a heavy building paper. It is used under steel roofing because acids in tar- or asphalt-saturated felts corrode the metal.

Roll Roofing. Good quality roll roofing that has been properly laid is a suitable low-cost first-cost covering for smaller farm buildings.

The roofing, which is also known as "prepared", "ready", and "composition" roofing, is composed of asphalt-saturated felt coated with asphalt. It is available in different grades or thicknesses. The heavier grades generally prove more satisfactory and give longer service.

There are three forms of roll roofing available. (1) Mineral-surfaced roofing is coated with mineral granules (ceramic-coated rock or crushed slate) on the weather side and dusted with talc or mica on the underside. It comes in a variety of colors; the color of the granules determine the color of the roofing. (2) Smooth-surfaced roofing is not coated; both sides are dusted with talc or mica. It comes in one color only. (3) Selvage-edge or wide-selvage roofing is coated with asphalt and mineral granules to 1 in. from the middle of the roll. The remainder is not coated with mineral granules, but there should be 2 or 3 in. of asphalt coating extending beyond the mineral granules for weather resistance where two strips of the roofing join. Because it provides two-ply coverage, selvage-edge roofing is more durable and more wind resistant than the other kinds of roll roofing and can be used on lower-pitched roofs. If you buy this roofing be sure that it is made for use with cold cement and that the cement and the roofing are made by the same manufacturer.

Following are installation details.

Deck. Roll roofing, both mineral surfaced and selvage edge, should be laid on tight sheathing.

Method. Roll roofing is usually laid with the sheets stretched parallel to the eaves. It can also be laid with the sheets stretched along the slope. If the latter method is used, fasten wood battens or metal strips over the long laps for more protection against tearing by the wind.

Mineral-surfaced roofing is laid as shown. Lap the strips 2 or 3 in. at the side or edge. Lap the ends of adjoining strips 4 to 6 in. Use large-headed galvanized nails to fasten the roofing. Space them 2 to 3 in. apart. If a nail goes into a crack between boards, pull it out and patch the hole in the roofing. Tin caps are not recommended. They corrode quickly and leave the nailhead protruding, which makes it easy for the wind to tear off the roofing.

Smooth-surfaced roofing may be laid in the same way as mineral-surfaced roofing; however, the "blind nailing"

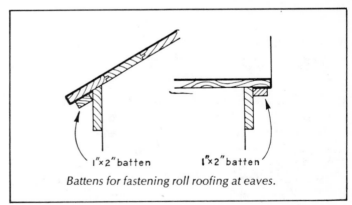

Battens for fastening roll roofing at eaves.

Roll roofing—average size and weight of commercial packages

Type	Width	Area per roll	Weight per roll
	Inches	Square feet	Pounds
Smooth surfaced...	36	108	55 to 90
Selvage edge...	36	108	70 to 74
Mineral surfaced...	36	108	45 to 65

Installation of regular roll roofing

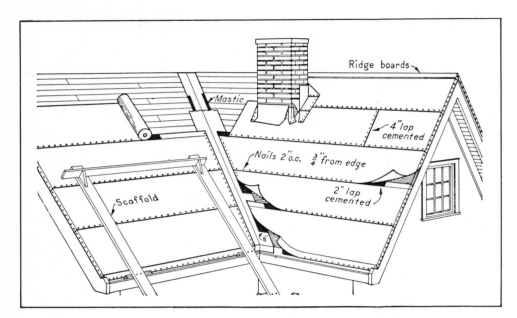

Installation of selvage-edge roll roofing

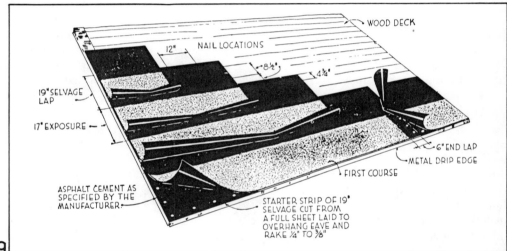

method is recommended. Increase the side lap to 4 in. and the end lap to 6 in. Nail the underlying edges through tin or fiber disks on 6-in. center. Cement the overlaying edges with hot asphalt or special blind-nailing cement. Step down firmly on these edges to make them stick.

Selvage-edge roofing is laid as shown. Be sure that no gap is left between the coated parts of adjoining strips. Sunlight will deteriorate any exposed, unprotected part of the roofing.

Roll roofing is usually fastened at eaves and gables by nailing into the edge of the sheathing. Shown is a better method, in which battens are used. Use barbed or cement-coated nails to fasten the battens.

Flashing. Flashings should be of the same material as the roofing and in two thicknesses. Rust-resistant metal chimney flashing should be used at chimneys with all but the cheapest roofing. Chimney flashing should be wedged and calked into the mortar joints.

Builtup Roofing

Builtup roofing consists of several layers of lightweight felt, lapped and cemented together with a bituminous material and covered with a layer of small-sized gravel or slag.

The roofing is long-lived and low-cost. It has high initial fire resistance, although it will burn freely once ignited.

The roofing may be used on roofs sloping ½ in. to 3 in. per ft. On greater slopes it may slip in hot weather and the gravel may not stay in place. On lesser slopes, the uneven surface may prevent proper drainage.

Builtup roofs may be 3-, 4-, or 5-ply, according to the number of layers of felt. A 5-ply roof, if laid by skilled workmen in accordance with the manufacturer's specifications, should last 20 years or more. Builtup roofs are usually installed by contractors who have the necessary equipment and experience.

Asphalt Shingles

Asphalt shingles, also called composition shingles, are

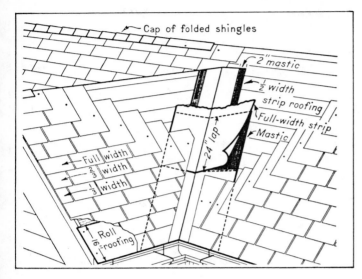

Installation of ordinary asphalt shingles. In general, valley construction details apply to all types of shingles.

Recommended minimum roof slope and approximate weight of various roof coverings

Type of roofing	Minimum rise per foot run with ordinary lap	Approximate weight per square [1]
	Inches	Pounds
Aluminum	4	30
Asbestos shingle:		
American multiple	5	300
American ranch	5	260
Asbestos, corrugated	3	300
Asphalt shingle:		
Lockdown	4	290
3-tab	4	210
Built-up roofing	½	600
Canvas (8 to 12 ounce)	½	25
Galvanized steel:		
Corrugated	4	100
V-crimp	2½	100
Roll roofing:		
Regular (2- to 4-inch lap)	4	100
Selvage edge (17- to 19-inch lap)	1	140
Slate	6	800
Tin:		
Standing seam	3	75
Flat seam	½	75
Wood shingles	6	200

[1] The different types of roofing vary in weight per square according to the weight or thickness of the roofing material itself.

widely used as a roof covering because of their moderate cost, light weight, and durability.

The shingles are composed of asphalt-saturated felt coated with asphalt and are surfaced with mineral granules on the weather side.

They are available as single shingles or in strips of several units, and in a wide variety of colors and patterns.

Asphalt shingles are semirigid and susceptible to damage by the wind. Also cheap shingles or shingles laid with too much surface exposed may curl badly after weathering. Some asphalt shingles are made so that they can be locked down or interlocked when laid. Strip shingles are available with a self-sealing compound on the tabs, and are especially useful in windy locations.

Strip shingles require less labor to apply than individual shingles. The three-tab strip shingle is one that is commonly used. It is 36 in. long and 12 in. high, and has cutouts 5 in. deep and ⅜ in. wide. These cutouts produce the appearance of individual shingles.

Shown is the general method of laying asphalt shingles. Detailed directions normally are included with the shingles when purchased.

Metal Roofing

Metal roofings include tin, galvanized steel, aluminum, copper, and zinc. Copper and zinc are not used much on farm buildings, because of the high cost. They are laid like tin roofing.

Metal roofings are light in weight and fire resistant. Those laid with locked or soldered joints can be used on low-pitched roofs with little danger of leakage.

Metal roofings have little insulating value, so insulating materials may be needed under them. Proper grounding, required for protection against lightning, is covered elsewhere in this chapter.

Tin

The so-called tin roofing is actually soft steel or wrought iron coated with a mixture of lead and tin. The material is more properly known as terne metal.

A tin roof of good material, properly laid and kept well painted, may last 40 to 50 years.

The roofing is available in strips 50 to 100 ft. in length and 14, 20, 24, and 28 in. in width. These strips come in rolls for easy handling.

The roofing is made in two thicknesses, IC and IX; IX is the heavier. It is available with a lead-tin coating of 8, 20, or 40 pounds per 436 square ft. Durability of the roofing depends on the thickness of the coating rather than on the thickness of the metal.

Special sheet-metal tools are required to lay tin roofing. If you are not experienced in laying the roofing, it may be advisable to have professional roofers do the work.

Following are installation details.

Deck. Tin roofing should be laid on tight sheathing. Tongue-and-groove boards are recommended. The boards should be well seasoned and of uniform thickness. The deck can be covered with rosin-sized or other tar-free sheathing paper to deaden the noise of wind and rain on the roof.

Method. If the roof slope is 3 in. or more per ft., a standing seam roof should be laid. If the slope is less than 3 in. per ft., a flat-seam roof should be laid.

Shown are details in laying a standing-seam roof. Form the seams as shown. The cleats should be of the same material as the roofing and should be spaced 8 to 12 in. apart. Fasten them securely to the sheathing. The finished seam should be straight, rounded neatly at the top edge, and

SECTION II: Setting Up the Basic Farm Structure

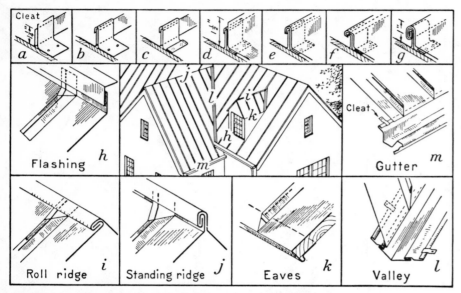

Installation of tin, or terne metal, roofing: a to g, steps in forming a standing seam; h to m, joints at breaks in the roof.

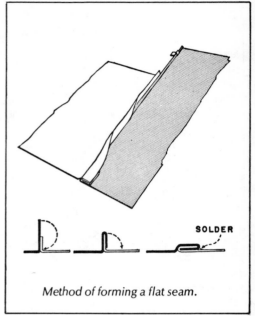

Method of forming a flat seam.

should stand 1 in. above the roof surface. Standing seams are not soldered.

A flat-seam roof is laid in the same general way as a standing-seam roof, except that the seams are formed differently and are flattened on the roof. Shown is the method of forming flat seams. These seams should be soldered to make them watertight.

Galvanized Steel

Galvanized steel is an economical and durable covering for farm buildings if good materials are used and if the roof is properly cared for.

Roofing made of alloy steel is more rust resistant than that made of plain steel, but durability depends chiefly on the protective zinc coating. Heavily galvanized roofing gives long service without painting. Lightly galvanized roof-

of quality". This is a heavy coating, and the roofing will last a long time under normal conditions. For maximum service, however, it will need painting eventually.

Galvanized steel roofing comes in different thicknesses, indicated by gage number. No. 28 gage or heavier is recommended for farm buildings. (The lower the gage number, the heavier the metal.)

Styles of galvanized-steel roofing commonly used on farm buildings are V-crimp sheets, corrugated sheets, and trapezoidal configurations of several shapes.

V-Crimp sheets are made to cover 24 or 30 in. allowing for side lap, and 6 to 12 ft. long. They come with 2, 3, or 5 V-crimps. The 5-crimp sheets provide a more watertight ing must be kept well painted after it begins to rust.

Roofing with a guaranteed minimum coating of 2 ounces of zinc per square foot is on the market under a special "seal.

Suggested Horizontal Projected Roof Lengths in Feet

FABRAL PROFILE	SIDELAPS	Pitch 2½" in 12"		Pitch 3" in 12"		Pitch 3½" in 12"		Pitch 4" in 12"	
		Endlap = 10"		Endlap = 9"		Endlap = 7"		Endlap = 6"	
		caulked	non-caulked (4)	caulked	non-caulked (4)	caulked	non-caulked (4)	caulked	non-caulked (4)
Grandrib		76	58	82	64	89	70	95	74
Prime Rib		59	46	65	51	70	55	75	59
Doublerib		57	44	63	49	68	53	73	56
2½" x ½"		44	—	48	37	52	40	55	43

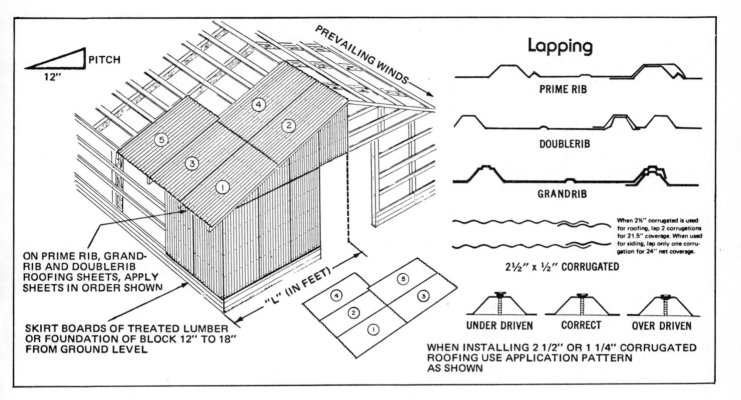

roof, because they are laid with a side lap of 2 crimps.

Corrugated sheets are 26 in. wide with 1½-in. corrugations, and 27½ in. wide with 2½ in. corrugations (24-in. coverage, in both cases, allowing for side lap). They are available in lengths of 6 to 32 ft.

Trapezoidal sheets are the strongest of the three and are available in the same widths, lengths, and thicknesses as the other two. Special sheets may cover 48 in. of width.

Following are installation details for sheet steel.

Deck. Corrugated and trapezoidal sheets may be laid on (1) tight sheathing; (2) 1-by-4 in. or 1-by-6 in. roofing slats, spaced 2 ft. apart, on top of rafters, spaced 2 ft. on center; (3) 2-by-4 in. purlins of top of rafters; or (4) rafters with 2-by-4 in. headers cut between them.

V-crimp sheets should be laid on tight sheathing. Tight sheathing under either type may be covered with rosin-sized sheathing paper.

Method. Corrugated sheets are laid with a side lap of 1½ corrugations. End lap is 9 in. if the roof slope is 4 in. rise per ft. run, and 6 in. if the slope is more than 6 in. rise per ft. run. All laps should be made over supports.

Fasten the sheets at all laps and intermediate supports. Nail down through the tops of the corrugation. Space the nails about 8 in. apart; screw-type or ring-shank nails are recommended. They should be long enough to fully penetrate the sheathing. The nails should have weather-protected heads, or neoprene washers should be used under the heads.

V-crimp sheets with 2 or 3 crimps should be laid with a side lap of 1 crimp. Those with 5 crimps should be laid with a side lap of 2 crimps. End lap and nailing is the same as for corrugated sheets.

Flashing. Flashings at hips, valleys, eaves and chimneys should be of galvanized steel. Open valleys should be used; they should be lined with galvanized steel 1 or 2 gages heavier than the roofing. Special valley sheets are available. You can also find special ridge rolls, joints, and flashings for use at hips, eaves, and chimneys; these will aid in making a tighter roof.

Painting. Galvanized-steel roofing should weather for at least a year before it is painted. Clean the roof thoroughly before you paint. Remove rusted spots with a wire brush. Remove any loose nails and renail. Do not paint unless the roof is absolutely dry; the best time is in warm, dry weather. When painting is needed, proper application of a good zinc-base paint will extend the life of the roof considerably.

Aluminum

Aluminum roofing is popular on farm buildings. Good grades of aluminum are highly corrosion resistant, require no painting, and have low maintenance cost.

Aluminum reflects a little more heat than steel. Reflection of the hot summer sun's heat keeps the interior of the building cooler.

Two styles of aluminum roofing commonly used on farm buildings are V-crimp sheets and corrugated sheets.

V-crimp sheets are made 26 in. wide with 5 crimps and about 50 in. wide with 8 crimps (24- and 48-in. coverage, respectively, allowing for side lap). In both widths they are made 6 to 32 ft. long, 0.019 in. and 0.024 in. thick, and have a smooth or embossed finish.

Corrugated sheets with either 1¼- or 2½-in. corruga-

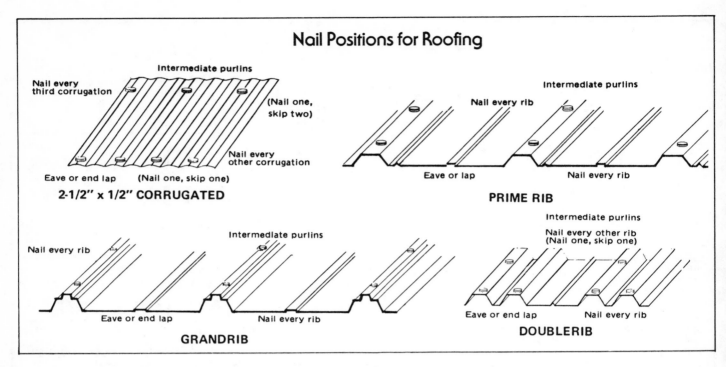

Aluminum Steel Roofing Load Chart—in Pounds per Square Foot

Profile	Gage	No. of Spans	Span in Inches *						
			18"	24"	30"	36"	42"	48"	54"
High Strength Grandrib	.0195	3 or More	188	106	67	46	34	26	—
Grandrib	.019	3 or more	138	78	49	34	24	19	—
Doublerib	.016	3 or More	102	58	37	26	19	—	—
2½ x ½	.019	3 or More	—	99	64	44	33	24	20

*Spans over 30" not recommended for roofing.

Galvanized Steel Roofing Load Chart—in Pounds per Square Foot
(Based on 80,000 P.S.I. TYP Yield Strength)

Profile	Gage	No. of Spans	Span In Inches						
			24"	30"	36"	42"	48"	54"	60"
Grandrib	29	2 Spans	141	89	62	44	34	27	21
		3 Spans	176	112	78	57	44	34	28
Prime Rib	29	2 Spans	97	60	41	29	21	—	—
		3 Spans	124	77	50	37	30	—	—
Doublerib	29	2 Spans	88	57	39	29	22	—	—
		3 Spans	110	71	49	36	27	—	—
2½ x ½	29	2 Spans	103	66	46	33	26	21	16
		3 Spans	123	79	55	40	31	24	20
	28	2 Spans	116	74	52	37	29	23	18
		3 Spans	139	89	62	45	35	27	22
1¼ x ¼	29	3 or More	31	19	14	10	—	—	—
	28	3 or More	34	22	15	11	—	—	—

tions are available in the same widths, lengths, and thicknesses as the V-crimp sheets.

Following are installation details for the corrugated sheets and V-crimp sheets.

Deck. Corrugated sheets may be laid on tight sheathing, 1-by-6 in. boards spaced 12 in. on center, or 2-by-4 in. purlins spaced a maximum of 24 in. on center. Spacing of the purlins will depend on the weight of the sheets and varies with corrugation size and thickness of sheet.

V-crimp sheets may be laid on tight sheathing or sheathing boards spaced up to 6 in. apart.

Tight sheathing under either type should be covered with 15-pound or heavier asphalt-saturated felt. Lay the felt in horizontal courses, starting at the eaves, and lap the courses a minimum of 3 in.

The roofing should not touch other kinds of metal. Cover steel nailheads in the sheathing (if not covered with asphalt felt) with asphalt-saturated felt or aluminum mastic. If metal purlins other than aluminum are used, coat them with aluminum-pigmented asphalt.

Method. Corrugated sheets with 2½ in. corrugations are laid with a side lap of 1½ corrugations. Sheets with 1¼-in. corrugations are laid with a side lap of 2½ in. corrugations. Side laps should be away from the prevailing winds. End laps should be 6 in. or more. At the eaves, extend the sheets 2 in. beyond the edge of the deck to form a drip edge. Fasten the sheets by nailing through the tops of the corrugations. Use aluminum nails, with neoprene washers under the heads, long enough to fully penetrate the sheathing.

V-crimp sheets—5 or 8 crimps—should be laid with a side lap of 2 crimps. End lap is 6 in. or more. Fasten the sheets by nailing down through the tops of the V-crimps. Use aluminum nails with neoprene washers under the heads.

Flashing. Aluminum flashing, 0.024 in. thick, should be used. Special ridge rolls, joints, and flashings are available for use at hips, eaves, side and end walls, valleys, and chimneys.

Canvas Roofing

Canvas roofing is light weight, long lived and watertight when kept well painted, and not hard to lay. It is particularly suitable for flat roofs that must be walked on, because it will not break readily.

The roofing is realtively high in cost, but is not expensive considering its durability. A good canvas roof, properly cared for, should last 25 to 30 years.

Canvas comes in two general classes—numbered duck and ounce duck. The difference is in the weaving. Numbered duck weighs from 7 to 20 ounces per linear yard 22 inches wide. No. 12 is the lightest and No. 00 is the heaviest. Ounce duck weighs from 6 to 15 ounces per linear yard 28¾ in. wide. It is made in three grades—army, double-filled, and single-filled. Of the three grades, only the army grade is suitable for roofing.

For roofing, canvas should be unbleached, unsized, closely woven and not lighter than the 10-ounce grade. No. 6 (13-ounce) and No. 2 (17-ounce) are recommended for roofs that must be walked on a great deal.

Canvas is available in widths up to 120 inches. However, the 22-in. width of numbered duck and the 28½-in. width of ounce duck are generally used for roofing.

Here are installation details.

Deck. Canvas must be laid on a smooth, tight surface. Tongue-and-groove flooring 2½ to 4 in. wide is recommended. Sheathing 6 to 8 in. wide could be used but is not as good because of greater shrinkage. If the boards cup or warp, the raised edges will make ridges that will wear through the canvas.

Method. There are several methods of laying canvas roofing. The following one is the most commonly used.

Paint the wood sheathing with a paint made of the following if a light-colored roof is desired: 100 pounds of white-lead paste, 4 gallons of raw linseed oil, 2 gallons of turpentine, and 1 pint of liquid drier.

When this paint is thoroughly dry, apply a heavy coat of the white-lead paste.

Lay the first strip of canvas on the wet paste and press it down firmly. (Pressing it down with rollers will give a smoother surface.) Stretch the canvas slightly and fasten it along the edges with ¾-in. copper tacks or galvanized nails, spaced 4 in. apart. Apply the white-lead paste along the edges. Lay the next strip of canvas with a lap of 1½ in. Nail the joint with ¾-in. copper tacks, spaced ¾ in. apart.

After the canvas has been laid, apply three coats of paint. For the priming coat, use the same mixture that was used to paint the sheathing, except use 3 gallons of raw linseed oil instead of 4. For the second and third coats use any good exterior paint.

Flashing. Canvas flashings should be used with canvas roofing. They are installed in the same way as are flashings of other materials.

Special Roofing Materials

Natural lighting, in addition to windows, may be desired in farm service buildings. Translucent structural panels made of plastic reinforced with glass fibers or wire and sold under various tradenames are available.

The panels are strong, light weight, durable, fire resistant, shatterproof, and require no painting. They are available in various colors. They admit soft, diffused light—up to 80% as much light as clear glass.

The panels are made in flat, corrugated, and V-crimp sheets. Corrugated sheets may be used with corrugated metal roofing or corrugated asbestos-cement roofing. V-crimp sheets may be used with 5-V-crimp metal roofing. The panels come in the same widths and lengths as the metal or asbestos-cement sheets.

Installation of the panels is the same as for metal or asbestos-cement sheets. Details may be obtained from manufacturers.

Flashing

Flashing, strips of metal or other material, must be installed in the valleys between intersecting roof surfaces and where the roof joins chimneys and other vertical surfaces to make the roof watertight.

Sheet metal—copper, aluminum, galvanized steel, or terne metal—is used for flashing with most types of roofing. Roll roofing or felt is used with bituminous roofings, and canvas with canvas roofing.

Special flashing materials are available. One type is steel sheets protected on both sides with baked-on coating or bonded asphalt-saturated fabric. Another type is a double layer of bituminous felt reinforced with cotton or steel-wire mesh.

Painting

Zinc, lead, and aluminum flashings are not ordinarily painted. Copper is sometimes painted to prevent the staining of other surfaces. Paint will not last long on untreated copper. Before painting, wash the surface of the copper with a solution of ½ gallon of lukewarm water, 4 ounces of copper sulfate, and ⅛ ounce of nitric acid. (Mix the solution in a glass container.) Wash the surface with water to remove all traces of the acid and allow it to dry.

Valley Flashing

Valleys may be open or closed. Open valleys should be at least 4 in. wide at the top and should widen out about ⅛ in. per ft. of length. Use flashing strips at least 20 in. wide.

When a valley is between roof surfaces of different areas of slopes, provide a baffle rib to prevent the larger or faster-descending volume of water from forcing its way up under the roofing on the opposite side. The baffle can be in the form of a V-crimp along the center line of the valley.

Shown are two methods of flashing closed valleys where rigid shingles are used. In the first, a continuous strip of metal is laid under the shingles. In the other, short pieces of metal are built in as the shingles are laid. If a prepunched nailhole in a slate or asbestos shingle falls over the metal flashing, provide a new hole. Each shingle should be fastened with two nails located outside the metal.

Vertical Flashing

Flashing at vertical surfaces, such as chimneys and walls, must extend up at least 6 in. and be counterflashed with cap flashing. The two flashings should not be fastened together rigidly.

The method of flashing a chimney located on the ridge is shown. The cap flashing should be built into the joints when the masonry is laid. It should be folded down at least 4 in. over the base flashing that is installed at the same time as the roofing.

Also shown is the method of flashing a chimney located on the slope. The cricket, or saddle, behind the chimney diverts water coming down the slope and prevents ice from forming behind the chimney.

Gutters and Downspouts

Gutters and downspouts are good investments, particularly in areas of high rainfall. They prevent the formation of water holes around buildings and damp conditions around foundations, and they reduce maintenance cost.

Gutters may be of wood built in as part of the cornice and lined with metal, or they may be metal troughs hung along the eaves.

Built-in gutters made of good materials will be more expensive than metal troughs, but maintenance cost will be considerably less. Built-in gutters should be built entirely outside the wall line of the building. The outer edge should slope to prevent breakage when ice forms in the gutter.

In areas of heavy snowfall, the outer edge of a gutter should be ½ in. below the extended edge of the roof. This is to prevent snowbanking on the edge of the roof and causing leaks. Hanging metal troughs are better adapted to such construction.

Gutters should slope 1/16 in. per ft. toward the outlet to the downspout. This outlet should be larger in circumference than the downspout. Cover the outlet with a wire guard to prevent the accumulation of leaves and trash in the downspout.

Downspouts must be large enough to remove the water from the gutter satisfactorily.

Conductor heads or funnels should be used where branch downspouts converge or at scupper of flat roofs.

Join sections of downspouts by fitting the upper section inside the lower section. Soldering the joints is not recommended. Solder the downspouts to the straps fastening them to the building.

On the lower end of a downspout, install:
(1) a shoe, or a turnout, if the water will merely drain away;
(2) a cast-iron or tile pipe connection or boot if the water will go into a storm sewer; or
(3) a rain switch, or diverter, if part of the water will go into a cistern.

Gutters and downspouts should be large enough to carry off normal storm water flow. Intense rains may occur periodically and the gutters overflow. However, if they overflow only for the duration of the storm, little damage will be done.

The table shows recommended sizes of downspouts and gutters for various roof areas. Local conditions may require larger sizes.

Install downspouts not more than 40 ft. apart.

Snow Guards

Snow guards should be used on steep roofs in cold cli-

Gutters and Downspouts

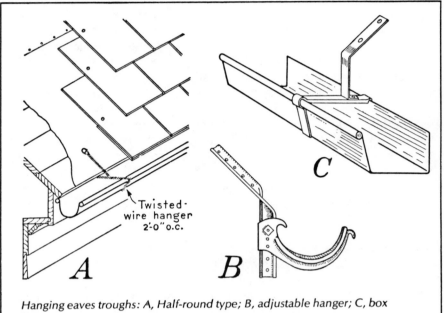

Hanging eaves troughs: A, Half-round type; B, adjustable hanger; C, box type.

Roof area	Gutter diameter	Downspout diameter
Square feet	Inches	Inches
100–800	4	3
800–1,000	5	3
1,000–1,400	5	4
1,400–2,000	6	4

Recommended sizes of gutters and downspouts for various roof areas.

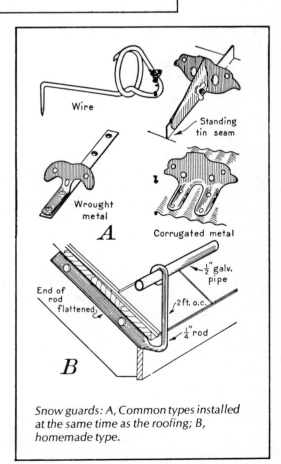

Snow guards: A, Common types installed at the same time as the roofing; B, homemade type.

mates to prevent sheets of ice or snow from sliding. Sliding ice can tear off roofing, break gutters, and endanger a person walking under the eaves.

Snow guards should be staggered in three rows near the eaves and spaced 6 to 12 in. apart. Sometimes, they are installed only over entrances or other traveled areas.

Roof Repair

Inspect the roofs of your buildings frequently. Check for breaks, missing shingles, choked gutters, damaged flashings, and defective mortar joints at chimneys, parapets, and coping.

Repair defects promptly. Don't neglect small defects. They extend rapidly and involve not only the roof covering but also the sheathing, framing, and interior finish.

You can probably repair small defects yourself. Hazardous repairs should be attempted by experienced persons. An inexperienced person can easily hurt himself.

Locating Leaks

As soon as a wet spot appears on a wall or ceiling, inspect the roof to determine the cause. The location of the spot may indicate the trouble. If it is near a chimney or exterior wall, look for defective or narrow flashing or loose mortar joints. On flat roofs, look also for choked downspouts or an accumulation of water or snow higher than the flashing. On sloping roofs, look also for corroded, loose, or displaced flashing and rotten shingles at valleys and at junction of dormers with the roof.

Other frequent causes of leaks are:
 (1) holes in the roof covering, generally the cause on plain roofs;
 (2) loose or defective flashing around cupolas and plumbing vent pipes;
 (3) gutters so arranged that when choked they overflow into the house;
 (4) a ridge of ice along the eaves that backs up melting snow under the shingles;
 (5) water leaking from downspouts that splashes against a wall and enters through a defect.

Shingles

Replace missing shingles with the same kind of shingle or a piece of rust-resistant metal. In an emergency, make a temporary repair with metal cut from a tin can.

If metal is used, paint it on both sides. Slip it under the shingle in the course above. Be careful not to dislodge sound shingles.

Metal Roofing

Close small holes in steel or tin roofing with a drop of solder. Solder a patch of the same kind of metal over large holes. If soldering tools are not handy, seal small holes with elastic roofer's cement. Paste a piece of canvas over large holes, using paint as the adhesive. Apply several coats of paint over the patch.

Close small holes in aluminum roofing with a sheet metal screw and neoprene washer or with an aluminum-pigmented caulking compound. Holes up to ⅜ in. in diameter can be closed also with cold solder. Holes over ½ in. in diameter should be covered with an aluminum patch. Coat the patch with aluminum-pigmented caulking mastic, and fasten it with sheet metal screws.

New short sheets may be used to repair large defects in metal roofing. If the defect is near the bottom of the old sheet, remove several fasteners, slip the new sheet under the damaged area, and refasten the old sheet in the same holes. If the defect is near the top, follow the same procedure, but place the new sheet over the damaged area.

Metal roofing may become riddled with small holes. An application of a heavy-brushing bituminous coating may be effective.

Flashing

Repairs to flashing should be made at the time the roofing is repaired or when inspection shows defects.

Fasten loose flashing securely in place and fill the joint with roofer's cement. If the joint is wide, oakum rolled in roofer's cement may be caulked in the joint.

Replace badly corroded metal in open valleys. Closed valleys are harder to repair. Where leaks occur, try to slip a piece of metal—a square piece folded on the diagonal—up under the shingles. If this cannot be done readily, call in a roofer.

Rake out loose mortar in chimney joints, and repoint the joints with a mixture of 1 part portland cement, 1 part lime, and 6 parts sand.

New Roofing Over Old

When you plan to reroof an old building, consider laying this new covering over the old. This is not always possible or desirable, but there are advantages.

The old roofing will provide additional insulation.

You can lay the new roofing without exposing the interior of the building or the sheathing to the weather.

You avoid the labor, expense, and mess of removing the old covering.

The roof framing must be strong enough to support the additional weight. If your roofing is exceptionally heavy, you may have to brace the rafters; or if they cannot be properly braced, you may have to remove the old covering.

Rigid shingles and metal roofings may be laid over old roll roofing and asphalt shingles, if the surface is not puffy or badly wrinkled. Puffy areas should be slit or cut and the old roofing nailed flat. If the new roofing is metal, cover the old roofing with rosin-sized paper (or asphalt-saturated felt for aluminum).

Metal roofings may be laid over old wood shingles. Nail

---

*For information on typical repairs, and which to try yourself, consult *Successful Roofing and Siding* by Robert C. Reschke.

2-by-4 in. nailing strips over the shingles, parallel to the eaves. Fasten the strips to the decking. For lightweight aluminum roofing, space the strips 16 in. on center. For steel roofing, space the strips 24 to 30 in. on center. End laps of the metal roofing sheets should be over strips. If the new roofing is aluminum, cover the nailheads in the strips with aluminum-pigmented mastic or asphalt-saturated felt.

New wood shingles may be laid over old. First, nail flat and secure all curled, badly warped, and loose shingles, and hammer down all protruding nails. Then follow these instructions:

(1) cut away 2 to 4 in. of the old shingles along eaves and gables;
(2) nail on wood strip to provide a firm nailing base;
(3) nail wood strip, level with the old shingle surface in open valleys and lay new metal valley sheet on top of wood strip;
(4) lay a double course of new shingles at eaves;
(5) install new chimney flashing;
(6) nail strips of bevel siding, thin edge down, at ridges to provide a solid nailing base.

Use five penny nails 1¾ in. long for the new shingles. The old shingles may have been laid on lath or strips. However, in nailing the new shingles, it is not necessary that the nails strike these strips.

8. Working With Corrugated Steel

Q. What is gage?
A. Gage is simply the measurement of the thickness of a galvanized steel sheet. The following gages and their decimal thicknesses represent the most popular roofing and siding products:

    29 gage — .0172"
    28 gage — .0187"
    26 gage — .0217"

There is only 15/10,000ths of an inch difference between 29 gage and 28 gage. This small thickness difference can mean no real strength between the two...yet, because it is thicker, 28 gage is more expensive.

Q. What is the strongest corrugated galvanized steel roofing and siding sheet available?
A. "Full-hard" steel, which means it is not annealed or heat-softened. Most producer's steel receives at least some annealing, which means that its grain structure (which is elongated and thus strengthened in the cold reducing process) is broken up and therefore weakened. Full-hard, high-tensile steel can be formed without any softening by means of special dies whch actually overform it and allow it to spring-back to pattern.

Q. Does the greater strength make it harder to nail?
A. Yes, but just a little. Remember, you are nailing through a gage or thickness that is too thin to present any real problem.

Q. Which is the better product...1¼ in. or 2½ in. corrugated?
A. The choice between the 1¼ in. and 2½ in. corrugation patterns is largely a matter of personal preference and, of course, habit or tradition in certain geographical areas. Actually, the 2½ in. x ½ in. pattern is stronger than the 1¼ in. x ¼ in. pattern because of its greater corrugation depth. The larger number of corrugations in the 1¼ in. corrugated sheet does not give it more strength.

Q. Do all metals rust?
A. Technically, no; but actually, yes. Technically, steel is the only metal that rusts because rust is defined as corroded or oxidized iron (steel). However, all other metals also corrode, which is caused by an instability in refined metals that tends to revert them back to their natural, unrefined states when attacked by moisture or moist air. Depending upon the metal, this corrosion or oxidation takes on various colors: steel's oxide is red, copper's oxide is light green, aluminum's oxide goes from white to grey to black. So by technical definition all metals do not rust, but in reality all metals do corrode or oxidize.

Q. What is the function of the zinc coating on a galvanized steel sheet?
A. The U.S. Bureau of Standards states that "Zinc is by far the best metallic coating for the protection of iron and steel against rust." Through the process of galvanizing, which moves a strip of clean steel through a bath of molten zinc, the zinc forms a tight metallurgical bond with the steel and greatly extends its service life. The heavier the zinc coating, the longer the service life. But more than just a skin-deep barrier, zinc also protects steel by electrochemical action. If a galvanized steel sheet is scratched or otherwise damaged and the zinc coating is broken, this electrochemical current fences the gaps and the zinc slowly sacrifices itself to continue its protection. This "healing" process is similar to the way human skin heals itself by scabbing. It is also the reason why you never see rust at the cut edge of a galvanized steel sheet, and why you never have to worry about cutting a sheet.

Q. Do heavier-gage galvanized steel sheets have a heavier zinc coating?
A. No, there is no more zinc coating on 26 gage sheets than on 28 gage sheets.

Q. What is G90 zinc coating?
A. G90 is a new zinc coating designation for galvanized steel sheets specified by the American Society for Testing and Materials and agreed to by American Iron & Steel Institute. G90 replaces the previous 1.25-ounce coating class, and refers to the fact that the old 1.25-ounce class required 0.90 ounce of zinc per square ft. total on both sides of the sheet by triple-spot test. All ASTM coating designations now bear the number of their triple-spot test weight, and sheets will bear this new designation in their brand marking.

Q. What is the 2-oz. "Seal of Quality" zinc coating?
A. 2-ounce "Seal of Quality" is a specification for a heavier zinc coating weight on galvanized steel sheets as defined by the American Zinc Institute. By triple-spot test, this specification requires 2.00 ounces of zinc per square foot total on both sides of the sheet. This "Seal of Quality" coating at slightly higher cost is intended for buildings in

areas of extremely rigorous weather conditions or highly corrosive industrial or seashore atmospheres.

Estimating Roofing and Siding Needs

The most accurate way to estimate galvanized steel sheets needed when ordering roofing and siding for a building is to determine the number of sheets needed for each surface of the structure and add those amounts together.

Estimating Roofing and Siding Costs

Roofing and siding is sold by the "square," or 100 sq. ft., of sheet area (without side and end lap allowance). To estimate the cost of roofing and siding for a farm building, mul-

Roofing

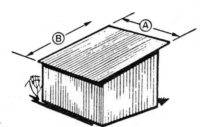

SHED TYPE

Divide distance "A" by 2 (24", or 2-foot, sheet cover width after side lap allowance) to get number of sheets per row. If distance "B" is 30' or less, one row of this sheet length is sufficient; if it is over 30', or if two rows are preferred, select two sheet lengths whose combined length equals distance "B" (depending upon slope, allow 6" to 8" end lap between rows; also add 3" overhang at eaves for drip edges). Multiply number of sheets per row by number of rows to determine total number of sheets. If different sheet lengths are needed, list number of each separately.

GAMBREL TYPE

Figure number of sheets for each surface, "A" and "B", by same as shed type. Add number of sheets for both "A" and "B" together and multiply by 2 to determine total number of sheets for both sides.

Siding

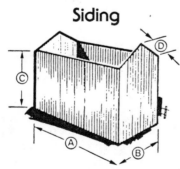

Divide distances "A" and "B" by 2 to get number of sheets for one side and one end, and multiply by 2 to get number of sheets for both sides and ends—all at distance "C" length. Divide distance "B" by 2 to get number of sheets for one gable end at distance "D" length, and multiply by 2 to get number of sheets for both gable ends. List number of each sheet length separately.

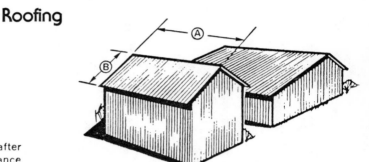

GABLE TYPE

Figure number of sheets for one surface same as shed type; multiply by 2 if both sides are same size, but add together if sides are different in size.

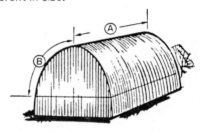

ARCH TYPE

Divide distance "A" by 2 to get number of sheets per row. Select sheet length or lengths equalling distance "B" (allow at least 9" end lap between rows). Multiply number of sheets per row by number of rows to determine total number of sheets.

Strongbarn Sheet Areas

Sheet Length (feet)	26" Wide (1¼" or 2½" corrugated)		27½" Wide (2½" corrugated)	
	Area per Sheet (sq. ft.)	Sheets per Square	Area per Sheet (sq. ft.)	Sheets per Square
6	13.000	7.692	13.750	7.273
7	15.167	6.593	16.042	6.234
8	17.333	5.769	18.333	5.455
9	19.500	5.128	20.625	4.849
10	21.667	4.615	22.917	4.364
11	23.833	4.196	25.208	3.967
12	26.000	3.846	27.500	3.636
13	28.167	3.550	29.792	3.357
14	30.333	3.297	32.083	3.117
15	32.500	3.077	34.375	2.909
16	34.667	2.885	36.667	2.727
17	36.833	2.715	38.958	2.567
18	39.000	2.564	41.250	2.424
19	41.167	2.429	43.542	2.297
20	43.333	2.308	45.833	2.182
21	45.500	2.198	48.125	2.078
22	47.667	2.098	50.417	1.984
23	49.833	2.007	52.708	1.897
24	52.000	1.923	55.000	1.818
25	54.167	1.846	57.292	1.745
26	56.333	1.775	59.583	1.678
27	58.500	1.709	61.875	1.616
28	60.667	1.648	64.167	1.558
39	62.833	1.592	66.458	1.505
30	65.000	1.539	68.750	1.455

Sheet Areas in Quantity

26" WIDE
(1¼" OR 2½" CORRUGATED)

AREA (square feet)

NUMBER OF SHEETS	SHEET LENGTH (feet)								
	6'	7'	8'	9'	10'	11'	12'	14'	16'
1	13	15	17	20	22	24	26	30	35
2	26	30	35	39	43	48	52	61	69
3	39	46	52	59	65	71	78	91	104
4	52	61	69	78	87	95	104	121	139
5	65	76	87	98	108	119	130	152	173
6	78	91	104	117	130	143	156	182	208
7	91	106	121	137	152	167	182	212	243
8	104	121	139	156	173	191	208	243	277
9	117	137	156	176	195	214	234	273	312
10	130	152	173	195	217	238	260	303	347
11	143	167	191	215	238	262	286	334	381
12	156	182	208	234	260	286	312	364	416
13	169	197	225	254	282	310	338	394	451
14	182	212	243	273	303	334	364	425	485
15	195	228	260	293	325	357	390	455	520
16	208	243	277	312	347	381	416	485	555
17	221	258	295	332	368	405	442	516	589
18	234	273	312	351	390	429	468	546	624
19	247	288	329	371	412	453	494	576	659
20	260	303	347	390	433	477	520	607	693
21	273	319	364	410	455	500	546	637	728
22	286	334	381	429	477	524	572	667	763
23	299	349	399	449	498	548	598	698	797
24	312	364	416	468	520	572	624	728	832
25	325	379	433	488	542	595	650	758	867
26	338	394	451	507	563	620	676	789	901
27	351	410	468	527	585	643	702	819	936
28	364	425	485	546	607	667	728	849	971
29	377	440	503	566	628	691	754	880	1005
30	390	455	520	585	650	715	780	910	1040
31	403	470	537	605	672	739	806	940	1075
32	416	485	555	624	693	763	832	971	1109
33	429	501	572	644	715	786	858	1001	1144
34	442	516	589	663	737	810	884	1031	1179
35	455	531	607	683	758	834	910	1062	1213
36	468	546	624	702	780	858	936	1092	1248
37	481	561	641	722	802	882	962	1122	1283
38	494	576	659	741	823	906	988	1153	1317
39	507	592	676	761	845	929	1014	1183	1352
40	520	607	693	780	867	953	1040	1213	1387
41	533	622	711	800	888	977	1066	1244	1421
42	546	637	728	819	910	1001	1092	1274	1456
43	559	652	745	839	932	1025	1118	1304	1491
44	572	667	763	853	953	1049	1144	1335	1525
45	585	683	780	878	975	1073	1170	1365	1560
46	598	698	797	897	997	1096	1196	1395	1595
47	611	713	815	917	1018	1120	1222	1426	1629
48	624	728	832	936	1040	1144	1248	1456	1664
49	637	743	849	956	1062	1168	1274	1486	1699
50	650	758	867	975	1083	1192	1300	1517	1733

27½" WIDE
(2½" CORRUGATED)

AREA (square feet)

NUMBER OF SHEETS	SHEET LENGTH (feet)								
	6'	7'	8'	9'	10'	11'	12'	14'	16'
1	14	16	18	21	23	25	28	32	37
2	28	32	37	41	46	50	55	64	73
3	41	48	55	62	69	76	83	96	110
4	55	64	73	83	92	101	110	128	147
5	69	80	92	108	115	126	138	160	183
6	83	96	110	124	138	151	165	192	220
7	96	112	128	144	160	176	193	225	257
8	110	128	147	165	183	202	220	257	293
9	124	144	165	186	206	227	248	289	330
10	138	160	183	206	229	252	275	321	367
11	151	176	202	227	252	277	303	353	403
12	165	193	220	248	275	302	330	385	440
13	179	209	238	268	298	328	358	417	477
14	193	225	257	289	321	353	385	449	513
15	206	241	275	309	344	378	413	481	550
16	220	257	293	330	367	403	440	513	587
17	234	272	312	351	390	429	468	545	623
18	248	289	330	371	413	454	495	577	660
19	261	305	348	392	435	479	523	610	697
20	275	321	367	413	458	504	550	642	733
21	289	337	385	433	481	529	578	674	770
22	303	353	403	454	504	555	605	706	807
23	316	369	422	474	527	580	633	738	843
24	330	385	440	495	550	605	660	770	880
25	344	401	458	516	573	630	688	802	917
26	358	417	477	536	596	655	715	834	953
27	371	433	495	557	619	681	743	866	990
28	385	449	513	578	642	706	770	898	1027
29	399	465	532	598	665	731	798	930	1063
30	413	481	550	619	688	756	825	962	1100
31	426	497	568	639	710	781	853	995	1167
32	440	513	587	660	733	807	880	1027	1173
33	454	529	605	681	756	832	908	1059	1210
34	468	545	623	701	779	857	935	1091	1247
35	481	561	642	722	802	882	963	1123	1283
36	495	578	660	743	825	907	990	1155	1320
37	509	594	678	763	848	933	1018	1187	1357
38	523	610	697	784	871	958	1045	1219	1393
39	536	626	715	804	894	983	1073	1251	1430
40	550	642	733	825	917	1008	1100	1283	1467
41	564	658	752	846	940	1034	1128	1315	1503
42	578	674	770	866	963	1059	1155	1347	1540
43	591	690	788	887	985	1084	1183	1380	1577
44	605	706	807	908	1008	1109	1210	1412	1613
45	619	722	825	928	1031	1134	1232	1444	1650
46	633	738	843	949	1054	1160	1265	1476	1687
47	646	754	862	969	1077	1185	1293	1508	1723
48	660	770	880	990	1100	1210	1320	1540	1760
49	674	786	898	1011	1123	1235	1348	1572	1797
50	688	802	917	1031	1146	1260	1375	1604	1833

For 13', 15' and 17' through 30' lengths, add areas of two given lengths.

EXAMPLES: One 13' long, 26" sheet: One 6' sheet = 13 sq. ft.
One 7' sheet = 15 sq. ft.
One 13' sheet = 28 sq. ft.

Two 30' long, 27½" sheets: Two 14' sheets = 64 sq. ft.
Two 16' sheets = 73 sq. ft.
Two 30' sheets = 137 sq. ft.

tiply the number of sheets needed for each surface by the area of each sheet length (see Sheet Areas Table). This will give a total sheet area in square feet. Divide this number by 100 to determine the number of squares needed. Multiply the number of squares by your dealer's price per square, and you have a close estimate of the roofing and siding portion of your total building cost.

For quick figuring, simply multiply the length by the width for each surface of the building and add these figures to get a total area in square feet. Divide this by 100 to get the number of squares needed, and multiply this by the price per square to get an estimate of your roofing and siding cost.

Storage and Usage

If sheets are going to be stored for any period of time, they should be unbundled and stood on end against an interior wall. They should be spread or fanned out at their bottom ends to allow for air circulation and, ideally, should stand on blocks.

If storage is to be for only a short time, sheets may be left in the bundle and covered with a cloth (not plastic) tarpaulin. Whether inside or out, blocks should be used to protect the bottom sheets from ground moisture.

When neither of these is followed and moisture forms

and remains trapped between sheets, whitish areas may appear. This is known as "wet storage stain," which is an oxide of the galvanized (zinc) coating that mars its appearance but does not decrease the service life of the sheets. For appearance's sake, be sure that sheets are free of "wet storage stain" at the time of purchase and store them properly if they are not to be used immediately.

Cutting

Sheets may be cut with sheet metal snips or with a power circular or jib saw equipped with a metal-cutting blade. When cutting with a power saw, keep the sheet lying flat and solid to avoid jumping or binding. It may also be cut lengthwise by scoring it with a pocket or utility knife and bending back on the score line.

Bending

Because of its high-tensile strength, sheets resist forming and should not be bent to a sharp corner. To bend a sheet around a corner, bend back on the peak of a corrugation, reversing it to get a long bend. Do not bend forward or down on the peak and thus close up the corrugation, as this creates a sharp, weak band that may break the sheet.

Fasteners

Nails

Hot-dipped galvanized steel ring- or screw-shank nails, with lead seals or lead, neoprene or rubber washers, are recommended for the application of steel roofing and siding. The galvanized or zinc-coated steel protects against rust and the seal or washer seals the nail hole and prevents leakage.

For best withdrawal resistance, select nail lengths for roofing and siding that are not so long as to go completely through the wood nailing strip, but long enough to allow a penetration of at least one inch. Following are nail length recommendations:

1¼ in. x ¼ in. corrugated—1¾ in. long nail for 2 x 4 roof purlin
1½ in. long nail for 2 x 4 side-wall girt

2½ in. x ½ in. corrugated—2 in. long nail for 2 x 4 roof purlin
1¾ in. long nail for 2 x 4 side-wall girt

Drill Screws

Roofing and siding may also be applied by the use of drill screws with sealing washers. They are self-drilling and self-seating with a power tool, and provide exceptional withdrawal resistance and a weathertight seal. These drill screws are available in plated carbon and stainless steel; since they are applied in the corrugation valleys, the shortest 1 in. length is recommended.

Sheet Metal Screws

Although not for use in fastening roofing and siding to the structure, galvanized sheet metal screws do come in handy and serve several important application functions:

(1) to draw roofing sheets tightly together at a side lap where misalignment has caused them to separate between nails;
(2) to draw roofing sheets tightly together at side laps when purlin spacing is 30 in. or more;
(3) to fill the hole where a nail has missed the nailing strip and must be pulled out.

Roofing Application

Slope

The minimum roof slope for steel sheets is 4 in. rise per ft., or 4:12. Apply a sealant at sheet side and end laps if less slope is used.

Purlin Spacing

The most practical roof purlin spacing is 24 in. center to center; however, some will safely span 30 in. to 36 in. (see load tables for specific spanning properties), but loading may increase deflection to the point of separating sheets at the side laps and necessitate the use of sheet metal screws to draw them back together. For best results, use 2 x 4 purlins on edge.

Procedure

The following steps pertain to 1¼ in.-corrugated, 26 in.-wide and 2½ in.-corrugated, 27½ in.-wide roofing; each pattern finishes one edge up and one edge down and covers 24 in.

(1) Start applying sheets vertically at the lower corner of the roof edge, downwind or away from the prevailing wind direction. Wind, rain and snow will thus blow over, not under, the sheet side laps.
(2) Use extra care in applying the first sheet. Be sure it is plumb, or perfectly square, with the roof edge so as to assure straight alignment of the entire row of sheets. One way to do this is to stretch chalklines between nails driven into each end of the ridge, and into the ends of each gable rafter to serve as guidelines.
(3) Start the first sheet with its edge down and extended at least one full corrugation over the gable end (unless a rake finishing accessory is used). Bend the sheet over and fasten it.
(4) Allow at least a 3-in.-sheet overhang at the eaves for a drip edge.
(5) If the roof slopes down 30 ft. long or less, only one horizontal row of sheets is needed; if it is longer, apply two rows by this method: Apply sheet #1 in the lower corner. Apply

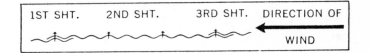

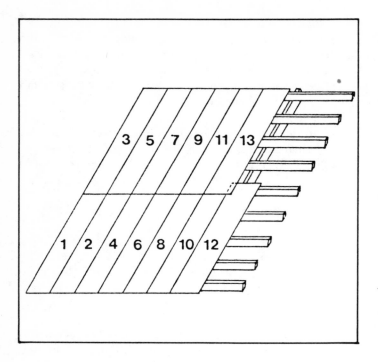

#2 next to and sidelapping over #1. Fasten sheets at the purlins, taking care to keep them in alignment at the top and bottom edges and at the side laps. Do *not* fasten the unlapped edges. Apply #3 above and end-lapping over #1 and the top corner of #2. Apply #4, side-lapping it over #2. Apply #5 above, end-lapping over #2 and the top corner of #4. Recheck the alignment. Repeat this procedure, always taking care to permanently fasten until all sheets are aligned and lapped.

(6) To keep labor at a minimum, stack sheets on a wagon or pickup truck; thus, as application progresses, the sheets may be moved easily.

Sheet Lapping Dimensions

Side laps for both patterns shorter than/longer than 30 ft. should be 1½ corrugations, leaving 24 in. covered.

End lap roofing 12 in. when roof slope is 3 in. (3:12) or less rise per foot (use sealant); 8 in. when roof slope is 4 in. (4:12) rise per foot; and 6 in. when roof slope is 5 in. (5:12) or more rise per foot.

Siding Application

Girt Spacing and Application

Wall girt or nailer spacing is 36 in. to 48 in.

Procedure. Following steps pertain to 1¼ in.-corrugated, 26-in.-wide and 2½ in.-corrugated, 26 in.-wide siding; the former finishes one edge up and one edge down and the latter finishes both edges down.

As with roofing, start applying sheets vertically at the wall corner downwind or away from the prevailing wind direction.

Again, use extra care in applying the first sheet. Be sure it is plumb, or perfectly square with the corner, so as to assure straight alignment of the entire row of sheets. One way to do this is to temporarily nail or "tack" a 2 x 4 (or a double angle trim accessory) horizontally at least 6 in. below the top of the baseboard, keeping it level or equi-distant with the rafter or truss plate, to serve as a guideline; rest the siding sheets on this strip, side lap, and fasten.

Start the first sheet, with its edge down, along the corner. Only one horizontal row of sheets is needed, as few farm building walls are more than 30 ft. high. Apply sheet #2, side-lapping it over #1, and repeat the procedure. Keep all sheets above the level of permanent contact with soil or manure.

Where it is necessary to notch the sheets to accommodate rafter ends, cover the notch area left exposed using a small scrap; this will make the building bird-proof.

As described earlier, it is possible to bend the sheets around a corner. However, like the roof rake, it is preferable to cut the sheet at a corner, start the remaining portion around the corner (which offers the opportunity for re-alignment), and cover the corner with a corner accessory.

For greater interior wall strength, such as that required for bulk crops storage, apply sheets horizontally up from the ground or floor by fastening them directly to the posts or studs.

Fastening

Roofing may be securely fastened by the use of ring- or screw-shank nails or drill screws, as follows.

Nails. Set the nail just off-center through the peak or top of the roofing corrugation; do not nail in the corrugation valley, which should be left free for efficient water run-off. Drive the nail at a slight angle to increase its holding power and to tighten the side lap, and drive it just far enough to set the sheet tight against the purlin. Do not over-drive, as this damages the washer, flattens the corrugation and causes misalignment. Do not underdrive, as this will result in loose application.

- In nailing through the four-sheet thickness created at the corner laps of a two-row roof, first use a nail punch or drill to ease the way; this may also be done for the two-sheet thicknesses of side and end laps.
- If a nail misses a purlin, pull it out and fill the hole with a sheet metal screw or sealant.
- Nail spacing: 1¼ in. corrugated, 26 in. wide roofing — every fifth corrugation peak, 126 per square at 24 in. purlin spacing; 2½ in.—corrugated, 27½ in. — wide roofing—every third corrugation peak, 96 per square at 24 in. purlin spacing.

Drill Screws. As opposed to nails, the drill screw is set vertically in the valley of the roofing corrugation and is driven until it sets the sheet tight against the purlin. However, at the side lap, the drill screw is set in the peak of the corrugation and driven into the purlin.

Recommended drill screw spacing is: 1¼ in.-corrugated, 26 in.-wide roofing — every fifth corrugation valley; 126 per square at 24 in. purlin spacing; 2½in-corrugated, 27½in.-

Load Tables

1¼" CORRUGATED STRONGBARN

ALLOWABLE LOADS & RESULTANT DEFLECTIONS

		PURLIN OR GIRT SPACING (Center to Center in Inches)											
		12	14	16	18	20	22	24	27	30	33	36	42	48
SINGLE SPAN	LOAD Lbs./sq. ft.	226	169	128	100	81	67	57	45	36	30	25	18	14
	DEFLECTION inches	.112	.156	.201	.252	.311	.377	.454	.574	.700	.854	1.008	1.344	1.784
DOUBLE SPAN	LOAD Lbs./sq. ft.	226	169	128	100	81	67	57	45	36	30	25	18	14
	DEFLECTION inches	.047	.065	.084	.105	.130	.157	.189	.239	.292	.356	.420	.560	.743
TRIPLE SPAN	LOAD Lbs./sq. ft.	283	211	160	126	101	84	71	56	45	37	31	23	18
	DEFLECTION inches	.075	.104	.134	.169	.207	.252	.302	.381	.467	.562	.666	.916	1.223

NOTE: Limiting steel design stress is 30,000 psi, determined in conformance to standard specifications as published in American Iron and Steel Institute "Light Gage Cold-Formed Steel Design Manual."

ALLOWABLE LOADS FOR DEFLECTIONS EQUAL TO 1/200TH SPAN

		PURLIN OR GIRT SPACING (Center to Center in Inches)											
		12	14	16	18	20	22	24	27	30	33	36	42	48
DEFLECTION		.060	.070	.080	.090	.100	.110	.120	.135	.150	.165	.180	.210	.240
SINGLE SPAN	LOAD Lbs./sq. ft.	120	76	51	36	26	20	15	11					
DOUBLE SPAN		226	169	122	86	63	47	36	25	19	14			
TRIPLE SPAN		226	142	95	67	49	37	28	20	14				

NOTE: Deflection equals purlin or girt spacing in inches divided by 200. Limiting steel design stress is equal to 30,000 psi, determined as stated above.

2½" CORRUGATED STRONGBARN

ALLOWABLE LOADS & RESULTANT DEFLECTIONS

		PURLIN OR GIRT SPACING (Center to Center in Inches)											
		18	21	24	27	30	33	36	39	42	45	48	54	60
SINGLE SPAN	LOAD Lbs./sq. ft.	198	146	112	88	71	59	50	42	36	32	28	22	18
	DEFLECTION inches	.129	.177	.231	.291	.358	.436	.523	.605	.697	.817	.925	1.165	1.452
DOUBLE SPAN	LOAD Lbs./sq. ft.	198	146	112	88	71	59	50	42	36	32	28	22	18
	DEFLECTION inches	.054	.073	.096	.121	.149	.181	.217	.251	.290	.339	.384	.484	.603
TRIPLE SPAN	LOAD Lbs./sq. ft.	248	182	139	110	89	74	62	53	46	40	35	28	22
	DEFLECTION inches	.086	.117	.152	.193	.238	.290	.344	.405	.473	.542	.614	.787	.942

NOTE: Limiting steel design stress is 30,000 psi, determined in conformance to standard specifications as published in American Iron and Steel Institute "Light Gage Cold-Formed Steel Design Manual."

ALLOWABLE LOADS FOR DEFLECTION EQUAL TO 1/200TH SPAN

		PURLIN OR GIRT SPACING (Center to Center in Inches)											
		18	21	24	27	30	33	36	39	42	45	48	54	60
DEFLECTION		.090	.105	.120	.135	.150	.165	.180	.195	.210	.225	.240	.270	.300
SINGLE SPAN	LOAD Lbs./sq. ft.	138	87	58	41	30	22	17						
DOUBLE SPAN		198	146	112	88	71	54	41	33	26				
TRIPLE SPAN		248	163	109	77	56	42	32	26	20				

NOTE: Deflection equals purlin or girt spacing in inches divided by 200. Limiting steel design stress is equal to 30,000 psi, determined as stated above.

wide roofing — every third corrugation valley; 96 per square at 24 in. purlin spacing.

Roofing Accessories

Use corrugated or plain ridge roll or cap to close the gap at the roof ridge, starting it downwind and end-lapping each section 4 in. or more. Fasten both edges of the ridge roll or cap over the roofing. The end wall flashing is necessary to cover the gap between a leanto roof and a building wall (apply the plain side under the building siding and the corrugated side over the leanto roofing); this same part may also be used to cover the joint of the gambrel roof by reversing or turning it over. The rake accessory is recommended to finish the roof gables.

Filler-Sealer Strips and Sealants

For best results in making a roof weathertight, use formed or plain rubber or polyurethane filler-sealer strips to fill in between the ridge roll or cap and the roofing, between the roofing and rafter or truss plate at the eaves, etc. Butyl rubber sealants are excellent for side lap and other sealing.

Skylights and Windows

Translucent fiberglass sheets are available for use as roof skylights and/or wall windows in a variety of colors and opacities.

Round or Arch Roof Application

Steel panels may be applied on round or Gothic arch roofs by following the recommendations below.

Minimum roof curvature radius is 16 ft; in other words, a 32 ft. building width is the narrowest round or arch roof to which steel panels may successfully be applied. If a smaller radius is used, it will be necessary to close the side lap gaps with sheet metal screws.

Maximum purlin spacing is 24 in. center-to-center to provide a continuous curved nailing surface and present the best appearance.

Apply the sheets vertically, and side lap both patterns 1½ corrugations; where more than one row of sheets is used, end lap 9 in.

Nail to the purlins through the corrugation peaks as on straight roofs; nail through every corrugation valley at both ends of each row of sheets so as to prevent the tension of the curved sheet from loosening the nails.

Drill screw both patterns through the valleys and side lap peaks as on straight roofs, and screw down every corrugation valley at both ends of each row of sheets.

Sheet Lapping

Side lap 1¼ in.-corrugated, 26 in.-wide siding 1½ corrugations to cover 24 in. Side lap 2½ in.-corrugated, 26 in.-wide siding 1 full corrugation to cover 24 in.

End lap siding 3 in., if necessary.

Fastening

As with roofing, siding may be fastened by the use of nails or drill screws, and the same general setting and driving recommendations prevail.

Nail spacing. 1¼ in.-corrugated, 26 in.-wide siding — every fifth corrugation peak, 91 per square at 36 in.-girt spacing; 2½ in.-corrugated, 26 in.-wide siding — every third corrugation peak, 69 per square at 36 in. girt spacing.

Drill screw spacing. 1¼ in.-corrugated, 26 in.-wide siding — every fifth corrugation valley, 91 per square at 36 in. girt spacing; 2½ in.-corrugated, 26 in.-wide siding — every third corrugation valley, 69 per square at 36 in. girt spacing.

Accessories

As recommended, use the double angle trim accessory to serve both as a guideline and a trim-closure for the siding. Also, use the corner accessory at the corners. Use the end cap, door track cover and other trim accessories to finish off all areas of the building. Application instructions come with these pieces.

Grounding

It has been estimated that 90 percent of farm buildings are not adequately grounded. Grounding protects against lightning damage and the possibility of fire, and its principle is to provide a path by which the electrical charge can enter the earth without passing through nonconducting parts of the building such as wood. A galvanized steel building may be easily grounded, as the roof itself forms the path from the ridge to the eaves and all that is needed to complete the path to the earth is a conductor. Follow these instructions to do the job properly.

1. On an average-length (200 feet or less in perimeter) building, install grounds at each of two diagonally opposite corners of the roof. On a very large building, install grounds at every 100 feet around its perimeter or at all four corners. Keep ground locations at least 6 ft. away from telephone or electric wires.

2. At each ground location, drive the ground pipe down into the earth to reach permanent moisture — usually 6 to 10 ft. deep. A 10-ft. length of ½ in. galvanized pipe, with 12 in. to 18 in. above ground, will do it.

3. Complete the path from the ground pipe to the eave with another ½ in. galvanized or with ⅜ in. steel cable. At least 18 in. of the pipe or cable should be in contact with the roof.

4. Fasten the connecting pipe or cable to the ground pipe with a cable clamp or two U-bolts.

5. Connect all metallic parts of the building, such as pipes or ventilators, to the grounding system. Make all connections secure by clamping or soldering, providing at least 3 in. of contact between two surfaces.

6. Lightning rods are not necessary, except where cupolas or other nonmetallic structures extend above the roof level. In such cases, run a rod 12 in. above the structure and connect it to the roof.

Improvements

Replacement

If an existing building has a good foundation and its structural framing is basically sound, the use of corrugated galvanized steel to replace its existing roofing and siding material is an economical way to make it look like new again and help it function usefully in the farm operation.

The most important consideration is the roof frame. Any sagging of the roof will loosen the roofing and cause leakage. Replace any wood members — roof rafters, purlins or decking, and wall studs and girts as well — that are cracked, rotted, split, warped, or broken. Reset all framing nails and drive additional ones where needed.

The next consideration is the existing roofing. If the material is asphalt or wood shingles or roll roofing and if most of it is damaged, it should be completely removed. But if most of it is in reasonably good shape, sheets may be applied right over it (be sure to use nails long enough to penetrate the wood decking through the extra thickness of shingles and roofing paper).

However, the ideal method is to apply sheets to 2 x 4 nailing strips fastened 24 in. to 30 in. apart over the shingles or roll roofing. In either case, leave as much of the existing roofing material on as possible, as it adds insulation value to the building.

If the existing roofing is corrugated galvanized steel, remove any badly damaged sheets and replace with the same pattern. Take care not to damage the usable sheets in the process of removing the nails.

If the existing roofing is aluminum and most of it is damaged, completely remove all sheets and replace. Do not combine the two materials because, besides presenting a different appearance, galvanized steel and aluminum are not metallically compatible and will have a corrosive effect upon each other.

Sealing Leaks

Leakage is seldom a problem with corrugated galvanized steel roofing. But when it does occur, and it is noticed at the side or end laps, simply reset all the nails to provide a tighter fit. As with new nailing, take care not to overdrive the nail to the extent of damaging the washer or denting the sheet. Use additional nails, where necessary, along the side and end laps.

If the nails will not hold tight enough and two sheets tend to separate at the side lap, use sheet metal screws to draw them tightly together. Punch or drill a small hole through both sheets at the corrugation peak and set the screw tight with a screwdriver.

Leakage through an enlarged nail hole in a sheet may be stopped by removing the old nail and filling the hole with a size 12 or 14 sheet metal screw. Then drive in a new nail about an inch or so away from the original hole.

Where water syphoning is taking place at a side lap, a sealant should be inserted between the sheets. Butyl rubber or a good caulking compound may be used (this type of sealant is also effective in sealing the roofing on low-rise roofs). To apply, first loosen the side lap nails with a screwdriver, pry up the overlapping sheet, force in the sealant in a steady bead the diameter of a pencil, and reset the nails; if the nails or washers become damaged or their holes become enlarged, follow recommended screw-sealing and renailing procedures.

If an extruded type of sealer is used to seal a side lap, be sure it is not less than 3/16 in. in diameter. Asbestos wicking may also be used as a seal. When the nails are reset or the screws drawn tight, all these materials form a gasket between two sheets to effectively seal their lap.

Painting

The life of galvanized steel roofing and siding on farm buildings varies according to climatic and atmospheric conditions, geographic areas, building usage, etc. But generally, it will give many years of rust-free service before it needs painting.

Whether your purpose is to lengthen its life or to change its appearance, follow these recommendations for the most satisfactory and economical painting.

Selection of paint is based upon recent research conducted by American Iron & Steel Institute, the trade association of basic steel producers. The A.I.S.I. study consisted of two years of outdoor field tests at three different locations and showed that both new and weathered galvanized steel may be painted with a minimum of preparation and with excellent adherence by following these basic guidelines:

(1) new galvanized steel may be painted immediately; weathering is desirable, but not necessary to achieve good adhesion when the correct paint is used;
(2) quality of paint is most important — purchase only top grades from reputable manufacturers;
(3) one paint coat can do the job, but two coats are better.

For best performance, use metallic zinc-dust paint in an oil, alkyd or phenolic base. Zinc-dust paints adhere best, will not peel, and provide long service life.

Next best are latex-base proprietary paints especially formulated for galvanized steel; these are well-suited for one-coat colored finish.

Cement-base paints provide good adhesion and are lowest in cost.

Some non-zinc-dust or non-cement base proprietary paints are satisfactory, but vary widely in adherence.

Do not use aluminum paint, as the two metals are incompatible and "bleeding" will occur.

Ordinary house or trim paints may be used as a second, or finish, coat over any recommended prime coat.

As mentioned, there is no need to wait before painting.

SECTION II: Setting Up the Basic Farm Structure

Likewise, it is not well to wait too long, or until the entire building is rusted, before painting old galvanized steel. The best time to paint is when the sheets first begin to show signs of rust.

Preparation for painting new or old galvanized steel is the same as for any other type painting.

(1) loosen rust or scale with a wire brush;
(2) remove loose dirt and rust as thoroughly as possible with a stiff broom;
(3) remove oil or grease with a solvent and wash entire surface with a deck scrub brush and water;
(4) apply paint only when the surface is dry, and in warm (at least 40° F.), dry weather.

Application of metallic zinc-dust paint or any of the other paints may be with a brush, a high-pressure sprayer or with a long nap, "industrial-type" roller.

Generally, the brush method provides the greatest coverage, but requires the most time and labor. The spray method requires the least amount of time and labor, but provides less coverage than brush. The roller method requires less time and labor than brush, but provides less coverage than spray. Any of these methods are satisfactory, so the choice depends upon you and what you figure your time is worth.

Projects Using Corrugated Steel

Following are some practical, do-it-yourself uses of corrugated galvanized steel sheets for other than complete buildings—applications that are useful on the farmstead or out in the field, and around and inside the farm home.

Roadside Stand
The sheets provide an inexpensive way to put up a permanent, all-weather roadside produce stand. Erect a simple wood frame (or use post-type construction) and apply sheets. Allow generous roof overhang. Paint white to reflect summer heat.

Forage Wagon
Instead of using expensive tongue-and-groove lumber for the sides and ends of the forage wagon, make it economical as well as useful by using low-cost galvanized steel sheets.

Wagon or Truck Sideboards
Bolt stakes to existing sides and nail sheets horizontally. Sideboards increase the carrying capacity of a farm wagon or truck.

Bin Liner
According to USDA estimates, one rat loose in a grain bin can eat or destroy $20 worth of grain in a year! Protect grain by lining the walls with steel sheets; rats can't gnaw through them. This also makes the bin easy to sweep clean for next filling.

FORAGE WAGON

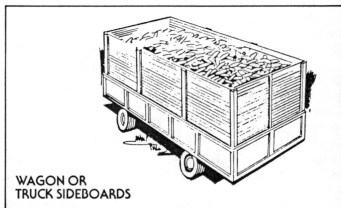
WAGON OR TRUCK SIDEBOARDS

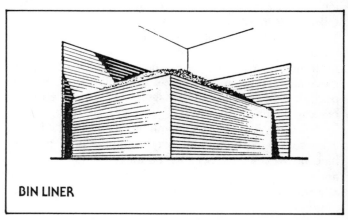
BIN LINER

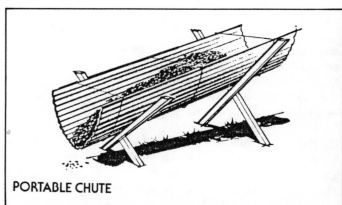
PORTABLE CHUTE

Working with Corrugated Steel

CATTLE FEEDER

PORTABLE HOG FEEDER

POULTRY HOUSE AWNINGS

10' x 18' CATTLE SHELTER

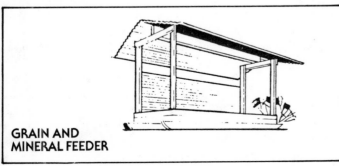
GRAIN AND MINERAL FEEDER

Portable Chute
A galvanized steel sheet, curved or bent into shape and secured with wire, makes a handy trough for unloading grain or ground feed into a bin or feeder. Chute is lightweight and easy to carry. Add axles and wheels for greater portability.

Cattle Feeder
Capacity depends upon the number of cattle. For every 100 bushels, build 8 ft. long, 6 ft. wide and 8 ft. high. For easy portability, mount skids on the feeder. Use sheets for the roof and center partition, and on the ends of windbreaks.

Portable Hog Feeder
Easy-to-build feeder holds 100 bushels of ground feed or shelled corn to feed over 40 large shoats. Erect simple wood frame, then cover with galvanized steel sheets. Gabled roof covers feeding space and helps protect swine and feed. Add skids for portability.

Poultry House Awnings
Cut sheets to size and hinge above all poultry house windows. Brace on the sides. Unless awnings are extra large, there is no need to construct frames; the rigid steel sheets hold its shape.

10 ft. x 18 ft. Cattle Shelter
Strong post-type cattle shelter is easy to erect. Sink four 14-ft. pressure-treated corner posts in ground to 3- or 4-ft. depth. Erect roof frame, using eight or more 2 x 4's attached to two 2 x 8's on each end. Secure frame to the posts and nail sheets across the top. Paint sheets white for maximum heat reflectivity.

Grain and Mineral Feeder
Some 2 x 4's and 1 x 4's and three sheets are all that's needed to make this efficient feeder. Compartments for grain and minerals are separate. Wide roof overhang keeps out weather. One sheet nailed to north side provides a windbreak to protect animals.

Portable Hog Shelter
This inexpensive, lightweight structure doubles as an individual farrowing house. Nail sheets to two cross-braced wood frames and join to a 2 x 4 ridge beam at top. Add skids for portability. Paint white for maximum heat reflectivity.

Portable Poultry Shelter
Basic materials for this 8 ft. x 10 ft. shelter are 20 lengths of 2x lumber, eight lengths of 1x lumber, poultry netting, welded wire fabric, and ten sheets. Light in weight, shelter can easily be towed from place to place.

8 ft. x 16 ft. Hog Shelter
Hog shelter is the same post type structure as cattle shelter, with shorter 10-ft. pressure-treated posts.

SECTION II: Setting Up the Basic Farm Structure

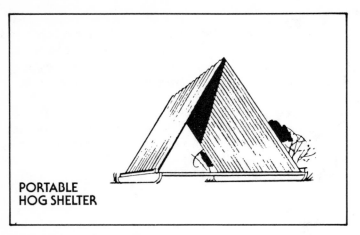
PORTABLE HOG SHELTER

PORTABLE POULTRY SHELTER

8' x 16' HOG SHELTER

PUMP HOUSE

Pump House
  This simple structure may be built by using some galvanized steel sheets and a few 2 x 4's. Sturdy, low-cost design assures trouble-free operation of the water supply system and protects the pump from weather, dirt, etc.

Protective Skirt For Raised Building
  Cut down on cold, damp floors and save on heat bills by nailing sheets to the exterior baseboard of a raised building. Or screw them to the bottom edge of a house trailer. Skirt also prevents trash from accumulating underneath and keeps rats and other animals away.

Portable Windbreak Snow Fence
  Sheets nailed horizontally to angle braces provide a strong, low-cost combination windbreak and snow fence. Height depends upon the severity of winter snows. Add skids for movement to any position on the farm lot or out in the fields near portable livestock shelters.

Storage Shelves
  Store feeds sacks, fertilizer, foodstuffs, luggage and other valuable items by building handy shelves of the sheets. The corrugated patterns permit air circulation, eliminates moisture beneath objects.

Screen and Window Storage
  Screens and storm windows fit into opposite-facing corrugation valleys of the galvanized steel sheets spaced proper distances apart.

Farm Building Plan Sources: your materials dealer; your County agent; Midwest Plan Service.
  Midwest Plan Service is a non-profit, self-supporting official activity of the following land grant universities and the U.S. Department of Agriculture:
  University of Alaska — College, Alaska 99645
  University of Illinois — Urbana, Illinois 61801
  Purdue University — Lafayette, Indiana 47907
  Iowa State University — Ames, Iowa 50010
  Kansas State University — Manhattan, Kansas 66504
  Michigan State University — East Lansing, Michigan 48823
  University of Minnesota — St. Paul, Minnesota 55101
  University of Missouri — Columbia, Missouri 65201
  University of Nebraska — Lincoln, Nebraska 68503
  Ohio State University — Columbus, Ohio 43210
  South Dakota State University — Brookings, S. Dakota 57006
  University of Wisconsin — Madison, Wisconsin 53706
  Write for a current Catalog of Books, Digests and Plans for Livestock Facilities and Storage Buildings. Plans may be purchased at nominal cost direct from the MWPS home office, from the Extension Agricultural Engineer at member universities, or from county extension offices.

PROTECTIVE SKIRT FOR RAISED BUILDING

PORTABLE WINDBREAK-SNOW FENCE

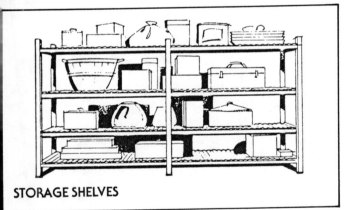
STORAGE SHELVES

SCREEN AND WINDOW STORAGE

State Universities. Most state universities with an Agricultural Engineering Department (other than the 13 members of the Midwest Plan Service) prepare and publish farm building plans and make them available at nominal cost. They may be obtained direct from the universities or through county agents.

U.S.D.A. Plans.

    Agricultural Engineer, Federal Extension Service
    U.S. Department of Agriculture
    Washington, D.C. 20250

USDA farm building plans are available from the agricultural engineering extensions of most state universities, from county extension offices, or direct from the above address. County agents have samples of all USDA and other plans, and can help in electing and securing them.

    Superintendent of Documents
    U.S. Government Printing Office
    Washington, D.C. 20402

Write for Price List 68—Farm Management, which lists several farm buildings and equipment plans available at nominal cost.

    American Society of Agricultural Engineers
    2950 Niles Road
    St. Joseph, Michigan 49085

Write for a list of independent consulting agricultural engineers who provide farm building planning services for a fee.

    Hoard's Dairyman Plan Service
    Fort Atkinson, Wisconsin 53538

A list of plans for free-stall and conventional dairy barns, calf and bull barns, dairy equipment, and other facilities—all available at nominal cost — is included in each issue of the publication.

    National Hog Farmer
    Swine Information Service
    1999 Shepard Road
    St. Paul, Minnesota 55116

A list of article reprints on swine housing and equipment — all available at nominal cost — is included in each issue of the publication.

9. Working With Plywood
(with Granular Storage Bin Specs)

Plywood Agricultural Construction Guide

Most farm buildings, by their very nature, are subject to hard usage and extreme moisture conditions. For this reason, agricultural structures—from animal barns to grain storage bins—must be designed for maximum durability and strength. For walls, doors, ceilings, roofs, and floors, plywood bearing the APA grade-trademark has many advantages. Low in-place cost, superior structural performance, durability, and minimum maintenance are just a few. This guide to plywood selection and usage in agricultural construction gives you basic information about technical design as well as building recommendations.

Plywood Defined

Plywood is a flat panel built up of sheets of veneer called plies. Veneers are produced by "peeling" a log; that is, placing it in a giant lathe and turning the log against a long knife. The veneer comes from the log in a long continuous ribbon that is cut into convenient size for kiln drying and assembly into plywood panels.

In manufacturing plywood, the dried veneers are assembled with the grain of each layer at right angles to the one next to it. Plywood always has an odd number of layers. A layer may consist of a single ply or two or more plies laminated with grain direction parallel. The layers are united with an adhesive under high pressure and temperature to produce a bond between the layers as strong as, or stronger than, the wood itself. Shown is typical plywood construction. Peeling veneers from a log and reassembling them also provides a means for making panels much larger than could be produced from solid wood. Standard plywood panels are 4 ft. by 8 ft., although wider and longer panels are also available.

Advantages

Strength. Wood is strong along the grain, but somewhat weaker across the grain. The cross-lamination of plywood manufacture distributes wood's along-the-grain strength in both directions, creating a panel that is very strong for its weight, and that resists splitting, puncturing, chipping, cracking-through, and crumbling. Plywood's two-way strength and its resulting high rigidity provides superior structural bracing.

Durability. Exterior-type plywood is bonded with a strong, 100 percent waterproof glue that will not weaken, even in boiling water. Silage acids, manure, urine, and other compounds common to the farm have no effect on the glue. Exterior-type plywood's high performance has been proven in all climates.

Plywood's in-place cost is low because it can be worked quickly and easily with common carpentry tools by anyone with ordinary skills. Economical to start with, the large panel size means time and labor savings. When buildings are designed to take advantage of plywood's structural properties, minimal framing is required. This, too, means added savings, and there is little waste with plywood; each panel covers 32 sq. ft.

Warmth. Like all real wood, plywood has excellent thermal insulation properties, which offer warm walls and ceilings to reduce condensation. Air won't seep through the panels as it does in other more porous construction materials. Plywood provides warmer, draft-free interiors because the joints are tighter and fewer in number.

Coolness. Throughout the summer plywood buildings are cool. The same thermal insulation properties that keep plywood buildings warm in winter also insulate them against the heat of summer. Painted white, plywood reflects heat better than aluminum. Unfinished, it reflects heat well, and as it ages its ability to reflect heat increases. Metals, on the other hand, tend to lose their heat reflection properties with age.

Convenience. Almost every lumber dealer in the nation sells plywood. Panels all of the same size stack and handle well in small, compact loads, which simplifies handling and use in all stages of construction. For example, a one foot stack of ⅜ in., 4 x 8 ft. plywood panels will contain enough material to cover 1,000 sq. ft. of roof or sidewall. Easily adapted to any framing system, plywood panels are true and square. Thus measuring, marking, and plumbing during construction are simplified.

Other Advantages. Plywood is strong, rigid, and weighs far less than most metals, lumber, or hardboard materials of equivalent strength. Dry from the mill, plywood is never "green." Its cross-laminated construction restricts expansion and contraction within the individual plies. From oven-dry to complete moisture saturation, a plywood panel swells across or along the grain only about 0.2 percent and considerably less with normal exposures.

Classification of Species

Working with Plywood

Group 1	Group 2		Group 3	Group 4	Group 5
Apitong [a][b]	Cedar, Port Orford	Maple, Black	Alder, Red	Aspen	Basswood
Beech, American	Cypress	Mengkulang [a]	Birch, Paper	Bigtooth	Fir, Balsam
Birch	Douglas Fir 2 [c]	Meranti, Red [a][d]	Cedar, Alaska	Quaking	Poplar, Balsam
Sweet	Fir	Mersawa [a]	Fir, Subalpine	Cativo	
Yellow	California Red	Pine	Hemlock, Eastern	Cedar	
Douglas Fir 1 [c]	Grand	Pond	Maple, Bigleaf	Incense	
Kapur [a]	Noble	Red	Pine	Western Red	
Keruing [a][b]	Pacific Silver	Virginia	Jack	Cottonwood	
Larch, Western	White	Western White	Lodgepole	Eastern	
Maple, Sugar	Hemlock, Western	Spruce	Ponderosa	Black (Western Poplar)	
Pine	Lauan	Red	Spruce	Pine	
Caribbean	Almon	Sitka	Redwood	Eastern White	
Ocote	Bagtikan	Sweetgum	Spruce	Sugar	
Pine, Southern	Mayapis	Tamarack	Black		
Loblolly	Red Lauan	Yellow Poplar	Engelmann		
Longleaf	Tangile		White		
Shortleaf	White Lauan				
Slash					
Tanoak					

(a) Each of these names represents a trade group of woods consisting of a number of closely related species.
(b) Species from the genus Dipterocarpus are marketed collectively. Apitong if originating in the Philippines; Keruing if originating in Malaysia or Indonesia.
(c) Douglas fir from trees grown in the states of Washington, Oregon, California, Idaho, Montana, Wyoming, and the Canadian Provinces of Alberta and British Columbia shall be classed as Douglas fir No. 1. Douglas fir from trees grown in the states of Nevada, Utah, Colorado, Arizona and New Mexico shall be classed as Douglas fir No. 2.
(d) Red Meranti shall be limited to species having a specific gravity of 0.41 or more based on green volume and oven dry weight.

Plywood Grades For Agricultural Applications

	Use these terms when you specify	Description and Most Common Uses	Typical Grade-trademarks	Veneer Grade Face	Veneer Grade Back	Veneer Grade Inner Plies	Most Common Thicknesses (inch) (3)
APPEARANCE (1)	A-A EXT-APA (2) (4)	Use where the appearance of both sides is important. Fences, built-ins, signs, cabinets, commercial refrigerators, tanks and ducts.	A-A·G-3·EXT-APA·PS 1·74	A	A	C	1/4, 3/8, 1/2, 5/8, 3/4
	A-B EXT-APA (2) (4)	For use similar to A-A EXT panels but where the appearance of one side is less important.	A-B·G-1·EXT-APA·PS 1·74	A	B	C	1/4, 3/8, 1/2, 5/8, 3/4
	A-C EXT-APA (2) (4)	Exterior use where the appearance of only one side is important. Sidings, soffits, fences, structural uses, truck lining and farm buildings. Tanks, commercial refrigerators.	A-C GROUP 2 EXTERIOR APA	A	C	C	1/4, 3/8, 1/2, 5/8, 3/4
	B-B EXT-APA (2) (4)	An outdoor utility panel with solid paintable faces for uses where higher quality is not necessary	BB·G-1·EXT-APA·PS 1·74	B	B	C	1/4, 3/8, 1/2, 5/8, 3/4
	B-C EXT-APA (2) (4)	An outdoor utility panel for farm service and work buildings, truck linings, containers, tanks, agricultural equipment.	B-C GROUP 3 EXTERIOR APA	B	C	C	1/4, 3/8, 1/2, 5/8, 3/4
ENGINEERED	C-D INT-APA w/ext glue (2)	A utility panel for use where exposure to weather and moisture will be limited.	C-D 32/16 INTERIOR APA EXTERIOR GLUE	C	D	D	5/16, 3/8, 1/2, 5/8, 3/4
	C-C EXT-APA (2)	Unsanded grade with waterproof bond for subflooring and roof decking, siding on service and farm buildings. Backing, crating, pallets and pallet bins.	C-C 32/16 EXTERIOR APA	C	C	C	5/16, 3/8, 1/2, 5/8, 3/4
	C-C PLUGGED EXT-APA (2)	For refrigerated or controlled atmosphere rooms. Also for pallets, fruit pallet bins, tanks, truck floors and linings. Touch-sanded.	C-C PLUGGED GROUP 4 EXTERIOR APA	C Plgd	C	C	1/2, 5/8, 3/4
	STRUCTURAL I & II C-D INT & C-C EXT-APA	For engineered applications in farm construction. Unsanded. For species requirements see (4).	STRUCTURAL I C-C 32/16 APA	C	C or D	C or D	5/16, 3/8, 1/2, 5/8, 3/4
SPECIALTY	HDO EXT-APA (2) (4)	Exterior type High Density Overlay plywood with hard, semi-opaque resin-fiber overlay. Abrasion resistant. Painting not ordinarily required. For concrete forms, signs, acid tanks, cabinets, counter tops and farm equipment.	HDO·A·A·G-1·EXT-APA·PS 1·74	A or B	A or B	C (5)	3/8, 1/2, 5/8, 3/4
	MDO EXT-APA (2) (4)	Exterior type Medium Density Overlay with smooth, opaque resin-fiber overlay heat-fused to one- or both panel faces. Ideal base for paint. Highly recommended for siding and other outdoor applications. Also good for built-ins and signs.	MDO·B·B·G-4·EXT-APA·PS 1·74	B	B or C	C	3/8, 1/2, 5/8, 3/4
	303 SIDING EXT-APA inc. Texture 1-11 (2) (7)	Grade designation covers proprietary plywood products for exterior siding, fencing, etc., with special surface treatment such as V-groove, channel groove, striated, brushed, rough-sawn.	303 SIDING 16 oc GROUP 1 EXTERIOR APA	(6) C	C	C	3/8, 1/2, 5/8, 3/4
	PLYRON EXT-APA (2)	Exterior panel surfaced both sides with hardboard for use in exterior applications. Faces are tempered, smooth or screened.	PLYRON EXT-APA PS 1·74			C	1/2, 5/8, 3/4

Notes:
(1) Sanded both sides except where decorative or other surfaces specified.
(2) Available in Group 1, 2, 3, 4, or 5 unless otherwise noted.
(3) Standard 4 × 8 panel sizes, other sizes available.
(4) Also available in STRUCTURAL I (all plies limited to Group 1 species) and II (limited to Groups 1, 2 and 3).
(5) Or C Plugged.
(6) C or better for 5-plies; C Plugged or better for 3-ply panels.
(7) Stud spacing is shown on grade stamp.

SECTION II: Setting Up the Basic Farm Structure

Plywood holds nails well and does not split when nails are driven close to the edges. Gluing and finishing plywood present no unusual problems. It may be sanded or textured, coated with a permanent finish, painted, stained, or left to weather naturally.

Selection and Use

Species Group

Some 70 species having varying strength and stiffness properties are used in plywood manufacture. To eliminate

APA Grade-Trademarks Explained

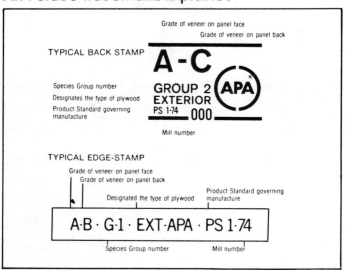

Veneer Grades

N Smooth surface "natural finish" veneer. Select all heartwood or all sapwood. Free of open defects. Allows not more than 6 repairs, wood only, per 4 x 8 panel, made parallel to grain and well matched for grain and color.

A Smooth, paintable. Not more than 18 neatly made repairs, boat, sled, or router type, and parallel to grain, permitted. May be used for natural finish in less demanding applications.

B Solid surface. Shims, circular repair plugs and tight knots to 1 inch across grain permitted. Some minor splits permitted.

C Tight knots to 1½ inch. Knotholes to 1 inch across grain and some to 1½ inch if total width of knots and knotholes is within specified limits. Synthetic or wood repairs. Discoloration and sanding defects that do not impair strength permitted. Limited splits allowed.

C Plugged Improved C veneer with splits limited to ⅛ inch width and knotholes and borer holes limited to ¼ x ½ inch. Admits some broken grain. Synthetic repairs permitted.

D Knots and knotholes to 2½ inch width across grain and ½ inch larger within specified limits. Limited splits are permitted.

Typical Engineered Grade Back-Stamp

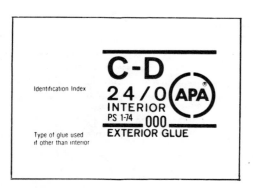

Identification Indexes for Unsanded Grades

Panels thicker than 7/8 inch shall be identified by group

Thickness (inch)	C-D INT - APA / C-C EXT - APA			NOTES:
	Group 1 & Structural I	Group 2 or 3 & Structural II*	Group 4**	
5/16	20/0	16/0	12/0	* Panels with Group 2 outer plies and special thickness and construction requirements, or STRUCTURAL II panels with Group 1 faces, may carry the Identification Index numbers shown for Group 1 panels.
3/8	24/0	20/0	16/0	
1/2	32/16	24/0	24/0	
5/8	42/20	32/16	30/12†	** Panels made with Group 4 outer plies may carry the Identification Index numbers shown for Group 3 panels when they conform to special thickness and construction requirements detailed in PS 1.
3/4	48/24	42/20	36/16†	
7/8	—	48/24	42/20	† Check local availability.

the need for individual species identification on panels, species meeting established standards are classified into five groups (see classification table). For comparison purposes, the lower the group number (Group 1 is lower than Group 4), the greater the stiffness and strength. When you see the word "Group", it refers to the species of the wood.

In accordance with U.S. Product Standard PS 1 for Construction and Industrial Plywood, the species of face and back plies may be from any group. However, when a face or back is made of more than one piece, the entire ply must be of the same species.

The group number appearing in the grade-trademark refers to the weakest species group used in either face or back, except for sanded or decorative panels ⅜ in. thick or less. These are identified by face species because their appearance is of primary importance.

Types of Plywood

Plywood produced under U.S. Product Standard PS 1 (the national industry manufacturing standard) is classified into two types—exterior and interior. In general, type is determined by the kind of adhesive plus the veneer grades used to make a plywood panel. These factors, along with proper manufacture, are the ones that largely control glue bond durability.

Exterior Type. Exterior plywood represents the ultimate in moisture resistance. A combination of waterproof glue and restrictive grading of veneers assures a plywood with a glue bond stronger and more durable than the wood itself. For permanent performance with exposure to moisture, always use exterior type.

Interior type. Interior plywood allows the use of either highly water-resistant or waterproof glues in combination with a lower veneer grade for inner plies and back than permitted in exterior type. It is suitable for use where there is a limited or controlled exposure to weather and moisture. For protected agricultural applications, interior plywood grades such as C-D INT with exterior glue, and STRUCTURAL I and II C-D are recommended.

Grades

Plywood grades recommended for agricultural use are shown in the table.

Appearance Grades. Within each type of plywood there are several appearance grades (as differentiated from engineered, or "workhorse" grades) determined by the grade of the veneer (N, A, B, C or D) used for the face and back of the panel. Panel grades are generally designated by type of glue and by veneer grade on the back and face. Grading characteristics are summarized in The Veneer Grades table. Details are contained in PS 1. Veneer grades and other information about the panels' makeup are shown in the grade-trademark.

Engineered Grades. These are unsanded grades especially designed for demanding construction applications where properties such as nail-holding, shear, compression, tension, etc., are of maximum importance. (Typical uses would be structural diaphragms, box beams, gusset plates, or stressed skin panels.) The engineered grades include C-D interior with exterior glue, STRUCTURAL I and II C-D Interior, STRUCTURAL I and II C-C Exterior, and C-C Exterior.

STRUCTURAL I and II grades are manufactured only with exterior glue. STRUCTURAL I grades contain only woods from Species Group 1, and STRUCTURAL II grades are limited to woods from Groups 1, 2, or 3. All unsanded engineered grades ⅞ in. thick or less carry an Identification Index number in the grade-trademark.

Identification Index. This is a pair of numbers separated by a slash in the APA grade-trademark on unsanded engineered grades.

The Identification Index tells you the panel's basic construction capabilities at a glance. The number to the left refers to the maximum recommended spacing in inches for supports when the panel is used for roof decking with face grain across supports. The number on the right refers to the recommended maximum spacing in inches for supports when the panel is used for residential subflooring with face grain across supports. (Example: An index number of $^{32}/_{16}$ means that the panel can be used for roof decking with supports spaced up to 32 in. o.c. and for subloors on supports spaced up to 16 in. o.c.) A number "0" on the right of the slash means the panel should not be used for subflooring. Wall sheathing should be specified in terms of thickness and Identification Index.

Identification Index and Farm Buildings. Variations with Building Codes: Building codes operating under the police power vested in a municipality are designed to ensure safety and health for both the occupants of the buildings and for the public. Code requirements for a structure vary somewhat depending on the type of occupancy; that is, whether theater, school, warehouse, or residence.

Some consideration should be given to the human occupany factors, even in structures that will be built outside of code jurisdiction. If the life hazard exposure is low, such as in crop storages, animal housing, or simple service buildings, greater support spacings may sometimes be used without reducing the structural adequacy of the building.

However, you should never use panels marked with identification index numbers lower than those shown in the tables herein for various support spacings.

For applications where structural loading is not really important, you have more latitude in selecting the grade and thickness you want. We show only minimum thicknesses in such applications. Within this limitation you can let appearance or cost dictate your choice of grade and thickness.

Use Recommendations

Exterior-type grade-trademarked plywood with a completely waterproof glue bond is recommended for construction of farm service buildings, both inside and outside. These buildings and the equipment used within them will invariably be subjected to high moisture conditions sometime during their lifetime. The C-C grade of exterior-type plywood is the most economical to use, with B-C or A-C grades as optional alternates.

In construction of commodity storages where the conditions are normally dry—moisture content less than 16 percent—C-D interior with exterior glue may be used as the structural liner. Where the moisture content is 16 percent or more, the plywood must be grade-trademarked C-C exterior, and panels must be applied so they are continuous over three or more spans. C-D interior with exterior glue may also be used for sheathing if it is not exposed to high moisture conditions or to the elements for an extended time.

Orientation. For maximum strength and stiffness in a building, apply plywood panels with the face grain perpendicular to the supports.

Spacing. When installing roof decks, wall sheathing or floor panels, follow this general rule for spacing: Leave 1/16 in. between panel ends and 1/8 in. between panel edges to allow for possible slight expansion. If wet or humid conditions are expected, double these spacings.

Some exceptions to the general rule are tongue-and-groove floors and certain wall panel installations. If a smooth, sanitary, draft-free floor surface is required, you may want to use tongue-and-groove plywood, preconditioned prior to application. Wall panels with shiplapped joints may be preconditioned in the same manner. In both cases, leave 1/16 in. space at panel ends and edges.

When panel joints are spaced you can minimize by starting application at the center and working toward each end.

Nailing. Use common nails for plywood: 6d or 5/16 in. through 1/2 inch panels; 8d for 5/8 in. through 7/8 in.; and 10d for 1 in. thickness. Space nails 6 in. o.c. at panel edges and 12 in. o.c. at intermediate supports except that when spans are 48 in. or more, nails should be 6 in. at all supports. Use galvanized or nonstaining nails to avoid rust streaks. For additional holding power, use ring-shank or screw-shank nails.

Selection. Select plywood thickness based on strength or appearance required, as given in tables and recommendations to follow.

Plywood Wall Construction

Sidings

Once limited primarily to single-family residential construction, plywood siding is more and more frequently appearing in various types of agricultural construction. Plywood sidings are extremely durable and strong. They are manufactured in a wide variety of surfaces: smooth and rough sawn, and groove patterns. The textured patterns generally are used for exterior surfaces, in both single- and double-wall construction, that are to be stain-finished. Use medium density overlaid (MDO) plywood for exterior and interior painted sur-

Typical Single Wall Applications

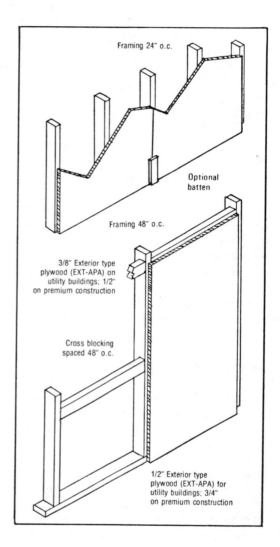

faces. This is plywood with resin-treated fiber surfaces fused to the panel faces under heat and pressure to produce an excellent base for paint and field-applied high-performance coatings.

Plywood sidings, finished with a qualified coating (carrying an American Plywood Association Certificate of Qualification), practically eliminate the need for periodic maintenance. Some of these coatings are factory-applied, others are designed for application in the field.

Single-Wall Construction

In single-wall construction, a single layer of plywood siding does the structural job of sheathing plus siding. Siding is applied directly to framing with no loss of strength, warmth or rigidity. Single-wall construction easily withstands the abuse of farm operations and pressure of wind and weather.

Minimum Plywood Recommendations For Single Wall Construction

Stud Spacing	Utility Buildings		Premium Construction		Nail Spacing (use nonstaining nails)	
	Plywood Thickness		Plywood Thickness		At Panel Edges	At Intermediate Supports
	Panels Vertical	Panels Horizontal	Panels Vertical	Panels Horizontal		
16" o.c.	5/16"	1/4"	3/8"	3/8"	6"	12"
24" o.c.	3/8"*	5/16"	1/2"	3/8"		
32" o.c.	**	3/8"	**	1/2"		
48" o.c.	1/2"‡	1/2"	3/4"‡	5/8"		

*For interior walls 1/4" sanded panels may be used over supports 24" o.c.
**Not modular to most common 48" panel width.
‡Cross blocking 4' o.c. between frames.

Minimum plywood thickness and installation recommendations are given. All panel edges should be placed over framing or blocking. 303 Texture 1-11 that is to be applied vertically should have cross-blocking at 32 in. o.c. when studs are spaced at 48 in. o.c. Leave 1/16 in. space at all panel and edge joints. When high humidity conditions are expected inside the building, space all panel joints to allow for expansion. End joint spacing in this case should be 1/8 in. and side joint spacing 1/4 in.

Typical single-wall constructions are shown. The siding joint details illustrated provide attractive as well as weather-tight joints between plywood panels.

Plywood-and-Pole Construction

Plywood is an excellent siding material for pole barn construction, easy to install with minimum labor, and capable of

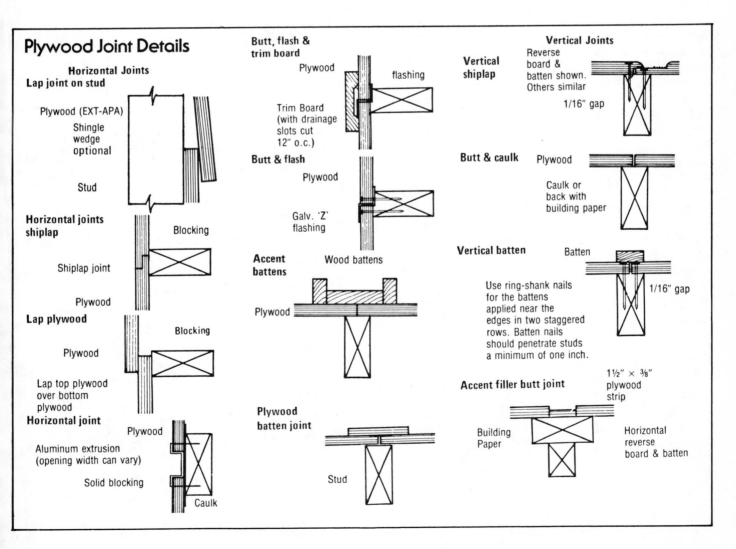

producing a tight wall with good structural characteristics.

For most farm service structures C-C exterior type plywood — the same structural workhorse panel recommended for dozens of other farm jobs— makes first-rate siding. It can produce a neat, trim, heavy-duty wall. However, you have other options. Plywood textured surfaces are worth considering where premium appearance is desired.

In most pole structures, framing girts are 2 in. dimension lumber, nailed flat to the poles. These girts, spaced 32 in. o.c., are adequate for ⅜ in. exterior type plywood siding. However, if the building is to be insulated, it may prove more convenient to space girts 24 in. to make use of economical batt insulation.

If insulation is not a factor, and minimum framing is desired, here's an option that may save money: Figure the job with horizontal girts spaced 4 ft; and with vertical blocking, also on 4 ft. centers, between the girts. This produces a grid framing pattern which is ideal for ½ in. exterior type plywood siding.

Use 6d common nails for ½ in. for thinner plywood; 8d nails for ⅝ in. and ¾ in. plywood. Nails should be spaced 6 in. around the panel edges and 12 in. at other supports. Blocking behind all panel edges provides a flat, tight weather seal. However, vertical joints can be covered with battens. Horizontal joints can be treated in a variety of ways. All joints should be spaced, rather than forced tight, to allow for expansion and contraction. End joints should be spaced about ⅛ in.; panel edge joints, ¼ in.

When trussed rafter spacing coincides with pole spacing, trusses may be bolted directly to the poles. It may be necessary to install two trussed rafters at each pole, one on each side of the pole, to adequately carry the roof load with wide spacing.

When it is inconvenient or impractical for pole spacing and rafter spacing to coincide, the best solution in most cases is a perimeter beam system. Fasten 2-in. members along a line at the tops of poles, forming support for trussed rafters. In installing such perimeter beams triangular exterior-type plywood gussets are recommended. The gib gussets provide a large area of contact for nailing, both in the 2-in. members

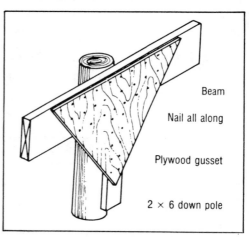

Typical Pole Construction

Plywood Gussets

Wall Insulation Values

	R-VALUE
Air film (15 mph wind)	0.17
3/8" EXT-APA Plywood	0.47
Air Spaces (3/4" to 4")	1.01
Air Film (still)	0.68
Blanket or Bat (approx. per inch)	3.10
Vermiculite (per inch)	2.13
Sawdust or Shavings (per inch)	2.22
Air Space (one reflective surface)	3.48
Expanded Polystyrene (per inch)	4.00
Polyurethane (per inch)	6.25

Typical Application Examples

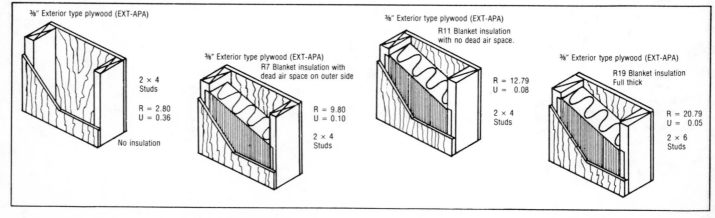

and in the pole. The gussets help transfer the roof load to the poles, and also form a tie across beam joints. Bracing the tops of the posts and the perimeter beam in this manner gives rack-resistant strength to the entire wall.

Insulated Wall Construction

Use is made of plywood's inherent insulating qualities when applied to both sides of the framing, with or without insulation in the dead-air space. You can prevent condensation within the wall by controlling inside humidity using adequate ventilation, and incorporating vapor barriers on the warm side of the wall. Shown is a typical insulated wall construction.

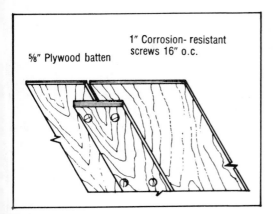

Plywood Ceiling Battens

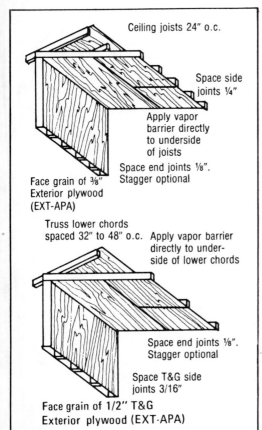

Plywood Ceilings

R values based on data for various types of insulation are given in Wall Insulation Values Table. R value refers to the resistance of a material to heat transfer, while the U value is a measure of its conductance, or the ease with which heat passes through an assembly of materials.

Plywood Ceilings

Typical recommended ceilings of grade-trademarked plywood are illustrated. Plywood is an excellent material for ceilings because it maintains its dimensional stability under conditions of extreme high humidity. It adds structurally also in that plywood can act as a brace for the bottom chords of roof trusses as well as support loads imposed by insulation installed on top of the ceiling deck.

Application Recommendations

When joints are spaced 24 in. o.c. or less, ⅜ in. minimum thickness of exterior type plywood is recommended. Apply with the face grain across the joints to obtain the maximum strength and stiffness from the plywood. No noticeable sag will occur if fill-type insulation is placed over the ceiling. Blocking at the panel edges will provide a smoother, dust-free ceiling, or plyclips may be used. Use one plyclip where supports are 24 in. o.c. or two for support spacings up to 48 in. o.c.

When high humidity conditions are expected inside of the building, space all joints to allow for expansion. End joint spacing should be ⅛ in.; side joint spacing, ¼ in.

Nail ⅜-in. plywood with 6d common nails spaced 6 in. o.c. along all panel edges over supports, and 12 in. o.c. over intermediate supports. Use galvanized or noncorrosive nails to avoid rusting.

Tongue-and-grooved ½ in. exterior type plywood is recommended as ceiling material over truss lower chords that are spaced 32 in. to 48 in. o.c. Alternates for T&G are square edge plywood with blocking at panel edges; or square edge plywood with plyclips, two per span.

Many poultry raisers have obtained satisfactory results with ⅜ in. Group 1, exterior plywood ceilings applied to trusses 32 in. or 48 in. o.c. However, it should be recognized that this spacing exceeds normal construction practice. In some applications, 2 x 4 battens applied over the joint from below have been used. The heavier batten is also convenient for mounting pipelines, lighting fixtures, etc.

Plywood battens may also be cut 3 in. to 4 in. in width from ⅝ in. exterior type (EXT) plywood and applied over ceiling joints as shown. Battens should be secured with 1 in. corrosion-resistant screws 16 in. o.c.

Plywood Roof Construction

Minimum Identification Index requirements for plywood roof sheathing are given in the Minimum Identification Index Requirements Table. Recommendations are based on use of C-C exterior plywood applied face grain across supports and will carry at least 40 psf total roof load. Check with your local

SECTION II: Setting Up the Basic Farm Structure

building inspection department for load conditions applicable in your area.

Rafters and Trusses

Plywood is unsurpassed as a roof sheathing material when applied with its face grain across the supports, because the panel's stiffness properties are fully utilized. The smooth, uniform surface of plywood is a superior base for built-up or roll roofing. And, wood or composition shingles benefit from plywood's high nail holding properties. Shown is a typical plywood-over-truss roof.

Differential deflection in abutting panel edges can result from unusual concentrated loads and cause roof damage. This can, however, be prevented with plyclips or blocking. As edge supports, plyclips contribute little, if any, to the stiffness of normal spans. But they do aid in preventing damage to the roof as the result of concentrated loads by maintaining alignment between panels.

Always support the panel edges where the following conditions occur:

(1) when ½ in. panels span more than 28 in.;
(2) when ⅝ in. panels span more than 32 in.;
(3) when ¾ in. panels exceed a 36-in. span.

If the support spacing is 48 in. or over, two plyclips are required. In roofing applications, design loads and deflection limits usually determine the panel thickness.

Plywood gussets can be used in the fabrication of trusses that vary greatly in span, slope, and design. Gussets and splice plates are nailed and glued in place to provide extremely strong, durable joints. Listed here are sources for trusses in spacings of 2, 4, 8, and 12 ft. with spans of 20 to 40 ft. and for roof slopes varying from 2 in 12 to 6 in 12 ft. For plans, write to: Agricultural Engineering Extension Service of Midwestern Land Grant Colleges and Universities, or Midwest Plan Service, Iowa State University, Ames, Iowa 50010; or, The Small Homes Council, One East Saint Mary's Road, Champaign, Illinois 61820.

Purlins

Shown is roof construction using plywood with purlins. For plywood thicknesses of ½ in. or less use 6d nails, over ½ in. use 8d nails. Space the nails 6 in. o.c. at panel edges and 12 in. o.c. at the intermediate supports. To assure a flat roof, leave 1/16 in. space between panel ends and ⅛ in. between edges as a precaution to allow for any possible expansion. If wet or humid conditions are expected, these spacings should be doubled.

Regardless of the roof framing system used, plywood is an excellent sheathing material. It sharply reduces material and labor waste, for the panels are precisely cut, sized, and squared. Its light wieght (about 1.1 psf for a ⅜ in. thick panel) makes using smaller supporting members possible. Strength and rigidity in a building are also greatly improved as the result of plywood's inherent resistance to racking.

Plywood Doors

Typical plywood doors are illustrated.

The plywood thickness required for a service door depends upon a number of factors: size of door, framing arrangement, wind load, method of handing, and use. Experience has shown that ⅜ in. sanded or 5/16 in. C-C exterior plywood are the minimum requirements for plywood "skins" over framing for farm building doors. For maximum stiffness and strength, the face grain of the plywood should run across the intermediate framing members.

C-C exterior (42/20 or 48/24) or B-C or A-C exterior (⅝ in. or ¾ in.) may be used as doors without framing—such as pen doors—and the size is normally limited to about 4 feet square. Large doors should use thinner plywood over framing so the door will remain flat.

Door Framing

Doors covered with plywood do not need diagonal bracing because the bracing action of the panel itself is excellent. Personnel-size doors, up to 4 x 8 ft., can be framed with cross-members as light as 1 x 4 lumber. Two center cross-members with a ⅜ in. plywood skin, or one cross-member

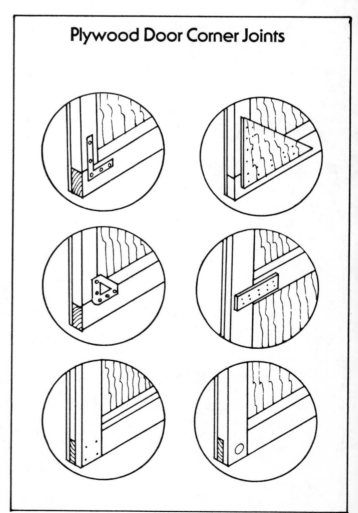

Plywood Door Corner Joints

with a ½ in. plywood thickness, normally are satisfactory. Large vehicle-size doors require heavier framing. Framing members placed on edged, rather than flat, are much stronger, but do make a thicker door. The plywood door skin, when well fastened, greatly strengthens all framing. A door with plywood on both faces is especially strong.

Corner Details

Doors with plywood on both faces have all framing joints held together by plywood's two way strength. Single skin doors must be built carefully to assure strength at the framing joints. Illustrated are some joint systems. The triangular inside corner gusset is especially strong.

Normally, 6d nails are used to fasten ½ in. plywood and thinner, and 8d for thicker. Nails should be spaced no more than 6 in. o.c. in all framing. Clinching the nails is desirable. Waterproof glues (resorcinol resin) give a very rigid door if proper application conditions can be followed— dry wood, warm temperature, plus pressure.

Insulated Doors

The natural insulating ability of plywood, especially in a double-skin door, provides a good insulation factor. Additional insulation may be required in doors constructed for cold storages. Plywood skins applied to the framework over sealing gaskets improve the performance of the door when rigid, batt-type, or fill insulation is installed within the door cavity.

Plywood Floors

Plywood floors are smoother because they have fewer joints. Construction economies are possible due to the range of panel thicknesses available. The Minimum Requirements Table gives recommendations for floors.

Always install plywood with face grain across supports for maximum strength. Blocking is recommended when a flat, strong floor is desired. Leave ⅛ in. spacing between panel edges and 1/16 in. at panel ends. When high humidity conditions are expected inside the building, space all joints to allow for expansion: end joints, ⅛ in.; side joints, ¼ in. Space the nails 6 in. o.c. around the panel edges and 10 in. o.c. at the intermediate supports. For plywood ½ in. thick or less, use 6d nails. When the supports are not of well-seasoned lumber, use ring-shank nails and set them 1/16 in.

Typical Plywood Door Construction

Minimum Identification Index Requirements for Plywood Roof Sheathing

Support Spacing (inches o.c.)	Identification Index	Panel Thickness (inch)
16	16/0	5/16 3/8
24	24/0	3/8 1/2
32	32/16*	1/2 5/8
48	48/24*	3/4 7/8

*Provide adequate blocking, tongue-and groove edges, or other edge support as Plyclips (1 for 32" spans. 2 for 48" spans).

SECTION III: Specialized Needs

Plywood Floors

Minimum Requirements For Plywood Floors

Support Spacing (inches o.c.)	Identification Index	Panel Thickness (inch)	Safe Load (psf)*	Safe Load (psf)**
12	24/0+	3/8 1/2	220	265
	32/16	1/2 5/8	335	435
	42/20	5/8 3/4 7/8	475	590
	48/24	3/4 7/8	675	780
16	24/0+	3/8 1/2	120	145
	32/16	1/2 5/8	185	240
	42/20	5/8 3/4 7/8	290	330
	48/24	3/4 7/8	380	440
24	32/16+	1/2 5/8	80	105
	42/20+	5/8 3/4 7/8	125	145
	48/24	3/4 7/8	165	190

* C-D INT w/exterior glue or C-C EXT-APA plywood. Face grain across supports; deflection limited to 1/180 of the span.
** If marked STRUCTURAL I.
+ Limited to light concentrated loads, such as foot traffic, and to uniform loads specified. For other conditions, specify greater Identification Index.

Minimum Identification Index For Shelled Corn or Wheat Bin Floors

Support Spacing (inches o.c.)	Level Depth of Fill (in feet, without surcharge)	Identification Index	Panel Thickness (inch)
12	6 and 8	32/16	1/2 5/8
	10 and 12	42/20	5/8 3/4 7/8
	16	48/24	3/4 7/8
16	6	42/20	5/8 3/4 7/8
	8	48/24	3/4 7/8
	10	48/24*	3/4

*If stamped STRUCTURAL I.

Minimum Identification Index For Shelled Corn or Wheat Bin Walls

Support Spacing (inches o.c.)	Level Depth of Fill (in feet, without surcharge)	Identification Index	Panel Thickness (inch)
12	8	12/0	5/16
	10	20/0	5/16 3/8
	12 and 16	24/0	3/8 1/2
	20	32/16	1/2 5/8
16	8	24/0	3/8 1/2
	10 and 12	32/16	1/2 5/8
	16 and 20	42/20	5/8 3/4 7/8
20	8	32/16	1/2 5/8
	10 and 12	42/20	5/8 3/4 7/8
	16 and 20	48/24	3/4 7/8
24	8	42/20	5/8 3/4 7/8
	10	48/24	3/4 7/8
	12	48/24*	3/4

*If stamped STRUCTURAL I.

Plywood Grain Storage Bins

Plywood's strength and stiffness properties make it an excellent material for construction of grain storage bins—portable, overhead, or hoppered. For details of design, see the "Granular Storage Design" section.

Floors

Since the weight of the stored grain can be considerable, selection of joist size and plywood identification index is critical. Minimum recommendations for shelled corn or wheat are given. Values are based on use of C-C EXT with face grain across supports and level fill of bin.

Side Walls

Plywood is used as an inside liner for grain storage structures and also as the exterior walls in either double-wall or single-wall construction. Again, the face grain direction of the plywood should run across the supports for maximum plywood strength. Adequate nailing is essential, particularly for single-wall construction. Deformed shank nails or ring shank nails will have a greater withdrawal resistance than common nails and are often used. Recommendations given are based on C-C EXT plywood, with loads based on level fill.

Fastening Plywood

Gluing

Any structural gluing in the erection of farm service buildings, such as in trusses, must be done with a completely waterproof glue. To date, resorcinol is the only glue that will meet the requirements, yet lend itself economically to field use.

Plywood's gluing characteristics are good; however, strong, durable joints are possible only when the glue manufacturer's directions are followed exactly. In gluing, special attention must be paid to the mixing of the glues, the spreading, the temperature and the pressure.

Resorcinol resin glue is dark in color, very strong, and completely waterproof. Sold in liquid form, it is accompanied by a powder catalyst which is added when the glue is mixed. Follow, very carefully, the glue manufacturer's directions and mix in small batches. Apply a thin coat to each member being glued, maintain close contact between members and allow 16 or more hours curing time. A room temperature of 70° or more must be observed during glue application and curing.

Casein glue should be used where only a slight resistance to moisture is required, for it is only moderately durable under damp conditions. Sold as a powder, it is mixed with water and applied by spreading the glue, joining the members and allowing time to cure. Follow closely the glue manufacturer's directions and recommendations.

Nailing

On any construction job, the cost of nails used is so small, compared with their importance, that they should always be of the best quality. Sizes (length) are indicated by "penny," abbreviated as "d" (as in 8d). Length of all nails will be the same in a particular penny size, regardless of thead or shank configuration. Only the diameter changes.

For long service and freedom from staining, nonstaining nails are recommended. The zinc coating on galvanized nails protects against rusting. Nails are also made of metals or alloys which are not subject to corrosion. Aluminum nails are a good example.

Common Nails. Common nails are for normal building construction conditions.

Box Nails. Smooth box nails of the same penny size will have a smaller diameter than common nails. Since this smaller diameter has less tendency to split the lumber, they are recommended for many uses. Though a box nail has less holding power than a common nail, it should be remembered that lumber split by a common nail is practically useless.

Cement Coated Nails. A frequent surface treatment for nails is "cement coating," which increases initial withdrawal resistance. This improved resistance diminishes with age, and the more dense woods show little benefit.

Pole Barn Nails. A special, thin, hardened, deformed shank nail has been developed to fasten framing to poles in pole buildings. These nails are made in sizes from 8d to 80d with 4 in., 5 in., and 6 in. lengths most commonly available.

Deformed Shank Nails. Various patterns of deformed shanks such as screw shank, ring shank, and barbed are available. These all have greater holding power than smooth nails and as a result deformed shank nails may often be smaller in size and still do the job satisfactorily.

Finishing Plywood

Initial Care and Finishing

Like any good finish material, plywood should be stored and handled with care to avoid damaging exposures before it has been finished. Storage should be in a cool dry place out of the sunlight and weather. If left outdoors, plywood stacks

Nailing

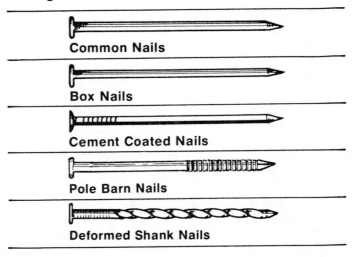

Basic Nail Information

Penny (d) Sizes: 2d, 3d, 4d, 6d, 8d, 10d, 12d, 16d, 20d, 30d, 40d, 50d, 60d

Nail Size and Number Per Pound

Size	Length (in.)	COMMON Diameter (in.)	COMMON No. per Pound	BOX Diameter (in.)	BOX No. per Pound
4d	1½	.102	316	.083	473
5d	1¾	.102	271	.083	406
6d	2	.115	181	.102	236
7d	2¼	.115	161	.102	210
8d	2½	.131	106	.115	145
10d	3	.148	69	.127	94
12d	3¼	.148	63	.127	88
16d	3½	.165	49	.134	71
20d	4	.203	31	.148	52
30d	4½	.220	24	.148	46
40d	5	.238	18	.165	35

should be covered in such a way as to provide good air circulation and ventilation between panels and to prevent moisture condensation and possible mold growth.

For best results, apply the first finish coat as soon as possible after the structure is complete.

Edge Sealing. All edges of plywood to be painted should be sealed with heavy application of exterior house paint primer or similar sealer. If plywood is not to be painted (e.g., finished with a stain), treat edges with a good water-repellent preservative compatible with the finish.

Sealing of all edges while the plywood is stacked is simple and easy. Horizontal edges, particularly lower drip edges of siding, should be treated with special care because of their greater wetting exposure.

Semitransparent or Penetrating Stain Finishes. High-quality oil base stains are recommended for textured plywood sidings, as these penetrate the wood surface and add color without film formation for a durable, breathing surface. Where maximum color and grain show-through are desired, the semitransparent stains should be used. One or two coats are applied in accordance with the manufacturer's directions, with two coats providing greater depth of color.

Best performance of stain finish comes from brushing on the stain. This tends to work the material into the wood surface and provides a uniform appearance. Application with a long napped roller is next best. Spray applications are not nearly as good and if used, should be followed by back brushing or wiping down to work the stain into the surface and under loose surface particles of saw-textured surfaces. Otherwise, the stain will come off with those particles during the natural weathering process.

Opaque or Heavy-Bodies Stain. If you want a solid even color, or if you want to hide all wood characteristics except texture, use oil-base opaque or highly pigmented stains. High-quality opaque oil stains penetrate the wood surface and provide a good bond to the panel. Don't use typical shake and shingle paints or stains because they don't penetrate the wood properly. Usually, within a short time, they crack and permit water to get underneath to flake away the finish.

Latex Stains. Available in both semitransparent and opaque types, emulsion stains appear to be durable and have excellent bond to plywood surfaces. Because they dry quickly, you must take care to avoid lap marks resulting from double coating.

Paint Finishes. Where paint must be used on textured surfaces, or where it is used on untextured plywood types, top-quality acrylic latex exterior house paint systems are recommended. A minimum two-coat paint system is essential, with the primer the more important always. A primer penetrates the wood to provide a good bond to the substrate and minimizes extractive staining. Some latex systems are designed to use an oil or alkyd primer, followed by the latex top coat. Others call for a specially formulated latex primer, usually with emulsified oil included. In any case, companion products, preferably made by the same manufacturer to be used together, are best.

To assure good penetration and bond, paint (especially the first coat) should be applied by brush. Use a stain-resistant primer with a compatible acrylic latex top coat. Two finish coats give much longer life and performance.

Checking may be expected, particularly on southern exposures, on painted Texture 1-11, 303 sidings and even with sanded plywood. Checks developed in textured surfaces generally are not objectionable since they blend with the architecturally rough surface. Even where checking occurs, bond and wear characteristics of a good quality latex paint have been found to be completely satisfactory. Where check-free surfaces are wanted, use Medium Density Overlaid Plywood.

Do not use clean film-forming finishes on plywood exposed outdoors.

Maintenance and Refinishing

The reason for finishing is to protect the surface of plywood siding and enhance its appearance. All finishes eventually deteriorate after extended exposure to sun and weather. For both stained and painted surfaces, appearance is the best guide as to when plywood siding should be refinished.

It is important that refinishing be done before the surface has weathered to a point where refinishing becomes a difficult and costly procedure, as well as ineffective. In the case of paint systems, too thick a film can quite often fail due to stresses set up within the film itself, causing it to crack and peel. Thus refinishing too often, before the film has weathered sufficiently, can be as bad as not maintaining the finish, and should be avoided.

Refinishing should always be done with top-quality finishes, in order to obtain optimum, long-term performance. The cost in labor for applying good finishes is no greater than for poor ones, and will be less at future refinishing time.

Surface Preparation. The amount of surface preparation needed depends on the type and condition of the original finish, as well as the material underneath.

For previously stained plywood, especially textured surfaces, you need only remove surface dust and finish chalk by hosing with a mild detergent solution, followed by a thorough rinsing with clear water. Then allow the surface to dry. In stubborn cases, scrubbing with a brush (nonmetallic) may be necessary. If surface fibers are obviously loosened or lifted, they should be removed first with mild bristle brushing before wet cleaning. Surfaces previously coated with an opaque stain may need a more vigorous cleaning, if the original film is at all loosened.

You can accomplish this with a water blaster, a high-pressure machine (generally available at equipment rental agencies) which often permits the use of hot water. When operating a water blaster, take care around window casings and siding joints to avoid forcing water into the wall.

For previously painted plywood, all loose paint, surface dirt and chalk must be removed. For textured plywood, loose paint usually can be removed satisfactorily with a stiff brush or, in more stubborn cases, with a water blaster. For sanded plywood, loose paint can usually be removed with moderate sanding, by scrubbing with a stiff-bristle brush, with power tools or with water blasting. When all loose paint has been eliminated, feather out edges of the remaining paint areas by sanding, to give a better, more even appearance for refinishing. Then remove all surface dirt and dust and allow to dry before painting. If paint deterioration is advanced, it may be best to remove all of the finish down to bare wood with power tools or use of a good chemical paint remover, preferably a water-rinsing type.

For severely checked surfaces, the checks should be filled before refinishing to provide an even surface for the new paint and to give best appearance and performance. Generally, it is difficult to fill checks with paint, and even where it is possible, the paint seldom remains intact. Face checking in plywood does not affect the siding's structural performance but only its appearance. Checking may occur when the plywood has been subjected to rapid moisture changes, particularly where panels are inadequately protected.

After cleanup described above, work a pliable patching compound into checks; let dry; sand smooth, and wipe clean.

Refinishing. Application of refinishing materials—stains or paints—is the same as for original finishing.

Reference Material

Plywood Properties

Nail Bearing. The density of plywood gives it exceptional nail-holding and nail-bearing properties.

Asphalt shingle nail tests indicate that the holding power of a 12-gauge nail in 5/16 in. plywood is great enough to cause tearing of the shingle before the nail withdraws.

In gusset design, lateral nail bearing is particularly important because it is the ability of the plywood to withstand loads that tend to pull a nail sideways. The Fastener Loads Table shows holding ability of various fasteners in plywood.

Vapor Transmission. Values given in the Vapor Transmission Rate Comparison Table represent water vapor transmission through plywood and other materials. Any material with a value below 1 permeance is considered a vapor barrier. The lower the value, the more efficient the barrier. Even unpainted ⅜ in. exterior plywood, with a permeance of 0.72, rates as a vapor barrier.

Thermal. Plywood, like all real wood, has excellent thermal insulating properties. Walls paneled with plywood and backed at all joints prevent drafts and air infiltration. A true dead air space is created, which acts as an insulating layer. Given also are insulating values for various plywood constructions.

Bending. Simple curves are easy to form with plywood. And the best results are obtained when a continuous rounded backing is used. Where the application calls for abrupt curvatures, secure the panel end to the shorter radius first.

In critical bends, two thin panels often work better than one thick one. Shorter radii can be developed by wetting and steaming (use only exterior-type plywood) but there is a greater risk of rupture, checking, and grain rise when this is done.

Average values in Minimum Bending Radii for Mill-Run Panels apply only for areas of clear straight grain. Occasionally, as the result of wood's natural growth characteristics—such as short grain—a panel may rupture at a longer radius than indicated. Where patches are present, minimum radii may be considerably longer.

Fire Resistance. A material's fire resistance is measured by the length of time it can contain a fire within a room or building. A one-hour rating, for example, means that the assembly will not collapse nor transmit flame or a high temperature while supporting its full load, for at least one hour after the fire commences. Given are specific plywood assemblies that have one-hour fire resistance. In addition, a roof of 1⅛ in. grade-trademarked 2.3.1. with exterior glue, T&G sides, on 4-in. wide beams spaced 4 ft. 0 in. o.c. qualifies for a "Heavy Timber" fire rating. Heavy Timber construction is markedly superior to most unprotected noncombustible (metal) structures under fire conditions.

Quantity Estimation - Storage

Bushels of grain or shelled corn in bin. Multiply the length by the width by the depth (all in feet) and divide by 1¼.

Bushels of ear corn in crib. Multiply the length by the width by the average depth (all in feet) and multiply by 2/5. If the crib is round, multiply the distance around the crib by the diameter by the depth of the corn (all in feet) and divide by 10.

Tons of hay in mow. Multiply the length by the width by the height (all in feet) and divide by 400 to 500, depending on the kind of hay and how long it has been in the mow.

Tons of hay in stack. Multiply the overthrow (the distance from the ground on one side over the top of the stack to the ground on the other side) by the length by the width (all in feet); multiply by 3; divide by 10 and then divide by 500 to 600.

Bushels in a container. Standard bushel used in the United States contains 2150.42 cu. in. Multiply bin length by width by height to find its capacity, then divide by cubic inches in a bushel.

Capacity of a cylindrical tank in gallons. Square the diameter in inches, multiply by the height in inches, and this product by 34. Move the decimal four places to the left and you have the capacity in gallons.

Lumber board feet. A board foot is the equivalent of a piece of lumber 1 in. thick, 12 in. wide, and 1 ft. long; in other words a sq. ft. of lumber 1 in. thick.

To find the number of board feet in a piece of lumber, multiply together: Nominal thickness in inches x Nominal width in inches x Length in feet and divide by 12; or multiply square feet of surface by thickness in inches.

Plywood and Support Bracing

Pressure (1) Zone	Identification (2) Index	Thickness of A-C EXT Group 1 (inch) (3)(4)	Support Spacing (inches)	Approximate Pressure (psf)
14	24/0	3/8	24	50
13	42/20	5/8	32	65
12	32/16	1/2	24	75
11	48/24	3/4	32	80
10	24/0	3/8	16	110
9	42/20	5/8	24	115
8	48/24	3/4	24	145
7	32/16	1/2	16	165
6	24/0	3/8	12	195
5	42/20	5/8	16	255
4	32/16	1/2	12	295
3	48/24	3/4	16	325
2	42/20	5/8	12	450
1	48/24	3/4	12	575

(1) For bin sizes where no number is shown, special design is required.
(2) For dry conditions (moisture content less than 16%), any sheathing grade may be used. Where moisture content is 16% or more, plywood must be C-C Exterior APA grade-trademarked, and continuous over 3 or more spans.
(3) For each step the group number of sanded plywood departs from Group 1, take one step down to the next smaller zone number. For example, if Group 3 plywood is used, and the applicable figure shows Zone 11, use the plywood and support spacing shown for Zone 9.
(4) Where moisture content is 16% or more, plywood must be continuous over 3 or more spans.

NAILS

Plywood Thickness (inch)	Ultimate Lateral Loads in Douglas fir lumber (lb. per common nail) *			
	6d	8d	10d	16d
5/16	275	305		
3/8	275	340		
1/2		350	425	
5/8		350	425	445
3/4			410	445

* Assume 3/8" edge distance.
For galvanized casing nails, multiply tabulated values by 0.6.

STAPLES

Penetration Into Lumber (inch)	Ultimate Loads * (lb. per staple)	
	Lateral Load	Withdrawal Load
3/4	160	100
1	180	150
1¼	200	200
1½	220	

* Values are for 3/8" and thicker plywood. 16 gauge galvanized staples with 7/16" crown, driven into Douglas fir lumber. Some plastic coated staples may provide higher values.

Working with Plywood

Other Design Considerations. In most bins the plywood will be installed on the inside of the wall framing. Consequently, the plywood will bear against the framing. In these cases, plywood should be nailed 6 in. on center at all edges and 12 in. on center at intermediate supports. Use 6d nails for plywood ½ in. thick and less, and 8d for thicker.

If the plywood is installed on the outside of the framing, the fasteners will be loaded in withdrawal and will require special design. In all cases and especially where the contents of the bin will accelerate corrosion, fasteners should be corrosion-resistant. In some, it may be desirable to "overdesign" the fasteners to allow for some losses due to corrosion. For design of fasteners in withdrawal, see the National Design Specification For Stress-Grade Lumber and its Fastenings (NDS).

The NDS should also be used when designing the wall framing. By applying the approximate pressures shown in Plywood and Support Bracing Table for each zone over the applicable portions of the wall, the designer can determine the structural requirements for the framing.

These recommendations are based on use of plywood that bears the grade-trademark of the American Plywood Association. For these highly stresses engineered applications, it is best to use plywood that meets the manufacturing standards of U.S. Product Standard PS 1 and the rigid performance requirements of the Association.

Plywood Insulation Values

	"R" Values**
⅜" Plywood (Single sheet)	.47
½" Plywood (Single sheet)	.62
⅝" Plywood (Single sheet)	.78
¾" Plywood (Single sheet)	.93

Plywood Inner Lining	Insulation Between Studs	Outer Wall	"U" Values*
¼"	None	⅜" Ext.	.38
¼"	R7 Blanket	⅜" Ext.	.10

*Coefficient of thermal transmission in B.T.U.'s/ hr./sq. ft./° Temp. diff.
**Thermal resistance, R = I/U

Fire Resistance

Location	Description	Rating
Roof or Floor	1⅛" APA grade-trademarked 2·4·1 with exterior glue, T&G sides, 4" wide beams (or two 2×'s) spaced to 4'-0" o.c. ⅛" galvanized metal furring strips 24" o.c., ⅝" Type X gypsumboard screwed to furring strips.	1-hour
Floor-Ceiling	⅝" APA grade-trademarked plywood UNDERLAYMENT—building paper—½" APA grade-trademarked C-D Interior plywood with exterior glue subfloor—2× joists spaced 16" o.c.—Acoustical tile suspended.	1-hour
Floor-Ceiling	23/32" APA grade-trademarked Group 1 UNDERLAYMENT with exterior glue, T&G sides—AFG-01 adhesive—2× joists spaced 24" o.c. (no bridging)—resilient furring channels 16" o.c.—⅝" gypsumboard ceiling—(6" wide × ⅝" gypsum wallboard stapled to underside of floor over plywood joints.)	1-hour
Exterior Wall	⅜" Exterior APA grade-trademarked panel or lap siding (includes nominal ⅜" 303 sidings)—½" gypsum sheathing—2 × 4 studs spaced 16" o.c.—⅝" Type X gypsumboard.	1-hour

Commodity Volume Weights
Hay, Bedding, and Feed Storage Space Requirements

Material	Weight Per Cubic Foot in Pounds	Cubic Feet Per Ton
Hay—Loose in shallow mows	4.0	512
Hay—Loose in deep mows	4.5	444
Hay—Baled loose	6	333
Hay—Baled tight	12	167
Hay—Chopped long cut	8	250
Hay—Chopped short cut	12	167
Straw—Loose	2-3	1000-667
Straw—Baled	4-6	500-333
Silage—Corn	35	57
Silage—Grass	40	50
Barley—48# 1 bu.	39	51
Corn, ear—70# 1 bu.	28	72
Corn, shelled—56# 1 bu.	45	44
Corn, cracked or corn meal—50# 1 bu.	40	50
Corn-and-cob meal—45# 1 bu.	36	56
Oats—32# 1 bu.	26	77
Oats, ground—22# 1 bu.	18	111
Oats, middlings—48# 1 bu.	39	51
Rye—56# 1 bu.	45	44
Wheat—60# 1 bu.	48	42
Soybeans—62# 1 bu.	50	40
Any small grain*	Use 4/5 of wt. of 1 bu.	
Most concentrates	45	44

*To determine space required for any small grain use wheat (60# = 1 bu.) for example. Then: 60 (4/5) = 48# wheat per cubic foot volume. To find number cubic feet wheat per ton, Then:

$$\frac{2000\# \text{ (Wt. of one ton)}}{48\# \text{ wheat per cubic foot volume}} = 42 \text{ cu. ft.}$$

10. Working with Adobe and Stabilized-Earth Block

Adobe and stabilized-earth blocks are inexpensive building materials for arid and semi-arid climates, although they have also proven satisfactory in humid areas when protected from free water. The map shows where they are generally used.

Adobe blocks are generally made of wet clay loam and straw, but with some soils the straw is omitted. Stabilized-earth blocks are made of sandy clay loam plus portland cement and water, or of sandy clay loam with a bituminous emulsion and water.

Earth-block buildings—that is, buildings built of either adobe blocks or stabilized-earth blocks—are desirable because: most of the building material is available at no cost; the buildings are strong, durable, and fire resistant; and, the massive walls maintain a comfortable temperature. There are disadvantages, however, such as: they deteriorate with long exposure to water; larger foundations are necessary because the walls are massive.

A factor that may be an advantage or a disadvantage is the cost of labor. Unskilled labor can be employed, but a lot of man-hours are required to make and lay the blocks.

Making the Blocks

Selecting the Soil

Select the soil for your earth blocks by the trial method. Start with a sandy clay loam—a soil that is neither high in clay content nor high in sand content. It should also be reasonably free of weeds, roots, and other organic matter.

Mix a sample block from the soil you have selected, and let it dry. If it warps or cracks when it dries, there is too much clay in the soil and you will have to mix sand with it to make a satisfactory building block. If the sample block crumbles, there is too much sand in the soil. You will have to add clay, or a stabilizer, to make a satisfactory block. The secret is to keep making sample blocks until you hit upon the right mixture.

Do not make earth blocks during freezing or rainy weather. Protect uncured blocks from frost; they will disintegrate if they freeze before they are cured.

Mixing the Soil

Prepare only as much soil at one time as you will need for one day's work. If the soil is cloddy, wet it the day before to soften the lumps.

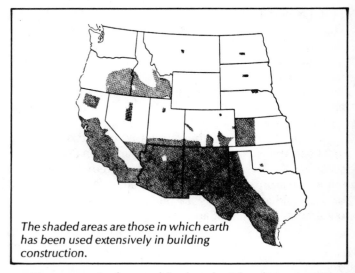

The shaded areas are those in which earth has been used extensively in building construction.

You can mix the mud by hand with a hoe, or with a machine—a hoe-type plaster mixer, a pug mill, or a dough mixer.

Adobe Blocks. Pile the soil in a 3- to 4-in. layer. Puddle it into a mud, and mix it thoroughly with a hoe. When it is uniformly wet, throw a layer of chopped straw on top and mix the straw into the mud. The layer of straw should be ¾ in. to 1 in. thick and the individual straws should be 2 to 6 in. long. If you mix the adobe in a machine, add 1 part straw to every 5 parts mud.

Be careful not to add too much straw. It will weaken the blocks.

Add water to the mud-straw mixture until the mixture is plastic enough to mold, yet stiff enough to pick up with a six-tined fork. It should be stiff enough to hold the shape of a block when the form is removed.

Stabilized Blocks. Portland cement and emulsified asphalt are the most common stabilizing additives. If you stabilize your blocks with asphalt, follow the directions of the asphalt-emulsion manufacturer. You can make blocks stabilized with portland cement by mixing soil and cement at a ratio of 1 part cement to 12 parts soil. More cement will make a stronger block; less cement, a weaker one. Then add water so that the mixture will form a block that can be handled, but will be at the same time as dry as possible. Too much water will reduce the strength of the cement. You can make 65 to 70 blocks, 4 x 6 x 12 in. each, from one bag of portland

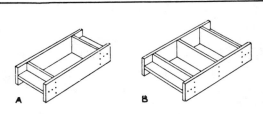

Forms used for molding earth blocks: A, Single form; B, double form.

cement mixed with soil at a ratio of 1 part cement to 12 parts soil. Cure all stabilized block by keeping damp for a week or two before drying.

Molding the Blocks

There are two ways to mold earth blocks—with a machine press or by casting the mud in forms by hand.

Several earth-block presses are on the market. The blocks made on them have two advantages over cast blocks: (1) the press-made blocks are more uniform in size and shape; (2) the press-made blocks are usually stronger, as much as twice as strong. But the presses make only one block at a time, and production is slower than by casting. Follow the press manufacturer's instructions; they will vary slightly with the press you buy.

Forms for molding cast blocks are of lightweight, surfaced wood or of metal. Make the inside dimensions the same as the block size you want. If you line the inside of the wooden forms with metal, the mud will not stick to them and blocks will be easier to remove.

The size of poured blocks most commonly made, and their approximate weights, are

4 x 8 x 16 in., 28 lbs.   5 x 12 x 16 in., 53 lbs.
4 x 10 x 16 in., 35 lbs.   5 x 10 x 20 in., 55 lbs.
4 x 9 x 18 in., 36 lbs.   5 x 12 x 18 in., 59 lbs.
4 x 12 x 18 in., 48 lbs.

Select a smooth, level area of ground for a molding site. If the ground does not have a good sod, scatter straw over it, or lay down heavy butcher paper or tar paper. This prevents the blocks from sticking to the ground. To mold the blocks:

(1) fork or shovel the prepared mud into the forms;
(2) press it into the forms with a tamper or with your hands, taking care to fill the corners of the forms;
(3) smooth the top of the mud with a stick or trowel;
(4) lift the forms up and away, and clean off the mud that sticks to them;
(5) repeat the process.

Two to four men working together can mix and mold 8 to 10 4 x 12 x 18 in. blocks per man-hour, each.

The number of blocks required to build 100 square feet of wall depends on the size of the exposed side of the block. For instance, when 4 x 10 x 16 in. blocks are laid in ½-in. mortar joints, 3 to 5 are needed for 100 sq. ft. of wall 16 in. thick, but only 195 blocks are needed for 100 sq. ft. of wall 10 in. thick.

A group of 3 men should be able to lay between 300 and 350 blocks in a wall in 8 hours.

Curing the Blocks

After the blocks have dried for a few days, stand them on edge so that both sides will have fairly equal exposure to the sun and wind. Let them dry this way for a week. When they are dry enough to handle, rub the loose dirt and straw from them. Stack them in a place where they will be protected from rain. When they have dried for 2 or 3 weeks in these stacks, you can build with them.

Laying the Blocks

Building Walls

Earth blocks are laid in a wall in much the same manner as ordinary burnt brick. Generally, mud without straw is used for mortar and the blocks are laid ½ to 1 in. in mortar joints. Lime mortar (1 part lime and 3 parts sand) or cement mortar (1 part portland cement and 2½ parts sand) is frequently used in permanent buildings. Lime or cement mortar costs more than mud, but it sets up faster and adds to the strength of the wall.

When the blocks are made of stabilized earth, stabilized earth is often used for mortar.

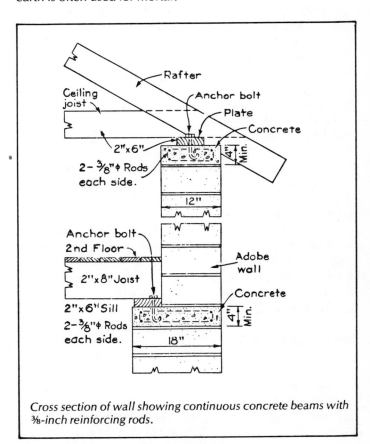

Cross section of wall showing continuous concrete beams with ⅜-inch reinforcing rods.

About 1 cu. ft. of mud or mortar is required to lay 15 to 17 blocks, 4 x 10 x 16 in. each, in ½-in. mortar joints.

The bearing walls of one-story adobe buildings and the second-story walls of two-story adobe buildings must be at least 12 in. thick. They should not be taller than 10 times their thickness. For example, a wall 12 in. thick should be no higher than 10 ft.

The lower wall of a two-story adobe building should be not less than 18 in. thick. Do not build adobe structures higher than two stories.

Stabilized-earth walls should not be taller than 12 times their thickness. And permanent buildings—such as houses—should not have walls taller than 8 times their thickness, whether built of adobe or stabilized earth, and they should not have unbuttressed walls longer than 20 times their thickness. Brace high or long walls until they have been permanently secured by plates and ceiling or floor joists.

Lintels are needed over door and window openings to support the wall above the opening, the roof rafters, and the second-floor joists. Make the lintels the same size and of the same material as you would for a burnt brick wall. Let them extend 9 to 12 in. beyond the jam on each side of the opening. Set them ½ to 1 in. higher than the window or door frame to allow for wall shrinking and settling.

For a permanent earth-block building, provide a continuous concrete beam (4 to 6 in. thick and as wide as the wall) under the floor and roof plates as shown. Reinforce these beams with two ⅜-in. steel rods on each side. The beams will distribute the floor and roof loads uniformly as well as stiffen and tie together the whole building.

Foundation. An earth-block building, like any other building, needs a good foundation. The foundation should be watertight concrete and should be at least 12 in. above the outside grade and 6 to 8 in. above a concrete floor. The top of the foundation should be dampproofed to prevent moisture from rising by capillary action from the ground into the wall. For details of foundation construction and dampproofing, see Farmers' Bulletin 1869, "Foundations for Farm Buildings." For a free copy, send a post card to the Office of Agriculture, Washington, D.C. 20250. Include your zip code.

Door and Window Frames

There are two ways to set door and window frames into an earth-block wall. You can build creosoted wooden blocks into each side of the opening and nail the door or window frames to them. The creosoted blocks should be 2 x 4's at least 12 in. long; there should be three of them built into each side of the opening. Or, you can set bolts in the wall and bolt a rough frame in the opening. Nail the finished frame to the rough frame when the wall has dried and settled.

Coating Outside Walls

Uncoated earth-block walls will last from 25 to 40 years in an arid climate if the top and the base are protected from moisture. An outside coating, however, will increase its life span. An outside coating is essential in a humid climate unless the blocks are well stabilized with cement or asphalt.

There are three types of outside-wall coatings:
(1) bituminous coatings;
(2) paint and whitewash coatings; and
(3) plaster coatings.

Bituminous Coatings. Hot tar, cold-pitch, asphalt, and Cunningham coal-tar paint are bituminous coatings.

Cunningham Coal-tar. Cunningham coal-tar paint is a mixture of 1 part portland cement, 1 part kerosene, and 4 parts coal tar by volume. The coal tar, also known as water-gas tar, can be obtained from local gas works or naval supply stores. It does not require heating or thinning with a solvent. Mix the cement and the kerosene first, and then stir them into the tar. If too thick, thin it with kerosene. Prime the wall with a thin coat of water-gas tar. Then apply the mixture with a brush or a swab.

Ordinary paints will not cover Cunningham coal-tar coating successfully. If you want to paint over a Cunningham coal-tar coating you will have to use asphalt-base aluminum paint as a primer coat.

Paint and Whitewash Coatings. Earth-block walls that do not have a bituminous coating can be painted. Linseed oil-lead paint is a durable satisfactory coating. Prime the earth blocks before you paint with linseed oil, or size them with a glue sizing. Make the sizing by mixing 1 pound of cheap glue sizing in 1 gallon of hot water. Thin the paint for the first coat, but apply it as it comes from the can for the second.

Whitewash is cheap and easily applied, but it is neither durable nor waterproof. You can make your own whitewash as follows:
(1) screen 50 lb. of hydrated lime into 6 gallons of water;
(2) let it stand overnight;
(3) strain out the lumps and foreign matter;
(4) thin to paint consistency with clean water.

You can make a longer-lasting, but more expensive, whitewash by soaking 5 lbs. of casein in 2 gallons of hot water until the casein is thoroughly softened (about 2 hours), and then dissolving 3 lbs. of TSP (trisodium phosphate) in 1 gallon of water. Add this solution to the casein and allow the mixture to dissolve. When the casein-TSP mixture is thoroughly cool, stir it into 8 gallons of cool lime paste. Make the lime paste by slaking 50 lbs. of hydrated lime in 6 gallons of water overnight. Just before using, dissolve 3 pints of formaldehyde in 3 gallons of clear water. Slowly add the formaldehyde solution to the casein-lime solution; stir constantly and vigorously. (If you add the formaldehyde too rapidly, the casein will jell and ruin the whitewash.) Mix enough for only one day's painting at a time; it does not keep.

Plaster Coatings. You can plaster outside walls with mud or with stucco. Mud plaster will improve the appearance of the building with little cost, but it must be painted to withstand the weather.

Mud plaster should be fairly stiff and fairly sandy. Mix 2 parts sand to 1 part mud. Apply it in two coats.

Lime-stucco and cement-stucco plasters are more durable

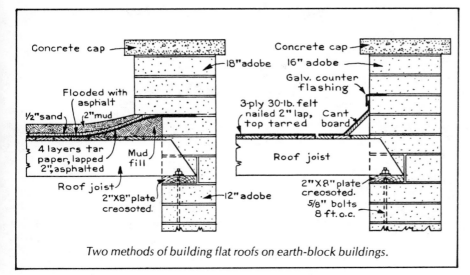

Two methods of building flat roofs on earth-block buildings.

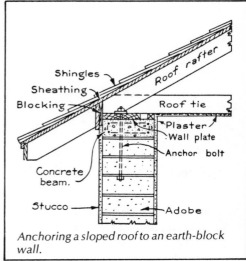

Anchoring a sloped roof to an earth-block wall.

than mud plaster. Cement stucco is more durable than lime stucco.

Allow the earth-block walls to dry and settle for 2 months before stuccoing them. Apply the stucco in two coats.

The first coat of stucco must be bonded to the wall. Illustrated is one method of bonding stucco with nails. Another method is to nail the first coat to the wall with tenpenny or twelvepenny nails. The nailing has to be done within 15 minutes after applying the first coat. Drive them flush with the mortar surface; space them about 12 in. at random (not in a straight line). A third method is to apply the stucco over metal lath.

The second coat of stucco has to be bonded to the first. The easiest way to make this bond is to scratch the first coat before it hardens. A board with nails driven through it, like a sharp rake, makes an excellent scratcher.

To make lime-stucco plaster, mix 1 part lime putty with 3 parts sand. Make the lime putty by slaking 44 pounds of hydrated lime or 27 pounds of quicklime in 6 gallons of water. Let the hydrated lime slake for at least 24 hours; let the quicklime slake for at least a week before mixing the plaster.

To make cement-stucco plaster, mix 1 part portland cement with about 3 parts sand. If you add 10 pounds of hydrated lime for each bag of cement the stucco will be easier to work.

Coating Inside Walls

Inside walls may be coated with paint or plaster like outside walls. You can plaster them first with mud and then with lime plaster, or finished off with paint, calcimine, or paper. They can be plastered with lime mortar over metal lath.

Animals like to lick and rub against earth walls. Protect the corners with corner boards and the door jambs with casings. Coat the interior walls that are within the reach of tied or penned animals with a bituminous coating or portland-cement plaster.

Roofing Earth-Block Buildings

An earth-block building can have almost any kind of roof, so long as the roof will keep rain water away from the earth-block walls.

In arid regions, flat roofs with parapets are popular. Two methods of building flat roofs are illustrated. Note that the top of the parapet is protected against deterioration with a concrete cap. Water is diverted from the roof-parapet joint in one case with a sloped mud fill, and in the other with a cant board (available in lumberyard) and flashing.

Outlet troughs (or scuppers) are necessary to drain water from flat roofs with parapets. Build them at least 3 ft. long so that they will dump the water away from the base of the wall.

You can make a good roof with felt and hot tar. Lay four or five layers of waterproof felt alternately with hot tar or asphalt. Cover the top with gravel, slag, or—in dry climates—earth.

In humid regions, the roof should be sloped and should have wide eaves. Farmers' Bulletin 2170, "Roofing Farm Buildings," describes roofs that can be adapted to earth-block buildings. The illustration here shows how to anchor a sloped roof to an earth-block building.

11. Interior Wiring

Components

Seriously consider contracting out the actual wiring of your building, since lack of experience can prove dangerous when installing wiring service. The basic principles and components are discussed below.

Conductors.
Electrical conductors provide paths for the flow of electric current and usually consist of copper or aluminum wire or cable over which an insulating material is formed. The insulating material insures that the current flows through the conductor rather than through extraneous paths such as conduits, water pipes, and so on. The wires or conductors are classified by type of insulation applied and wire gage; the insulation in turn is subdivided according to operating temperature and nature of use. Illustrations show the more common single conductors used in interior wiring systems.

Wire sizes. Wire sizes are indicated according to American Wire Gage (AWG) standards. The largest gage size is No. 0000. Wires larger than this are classified in size by their circular mil cross-sectional area. One circular mil is the area of a circle with a diameter of 1/1000 of an inch. Thus, if a wire has a diameter of 0.10 in. or 100 mil, the cross-sectional area is 100 x 100 or 10,000 circular mils. The most common wire sizes used in interior wiring are 14, 12, and 10; they are usually of solid construction.

Numbering. The higher the number, the smaller the size of the wire. The sizes most often used have even numbers, e.g., 14, 12, 10. Numbers 8 and 6 wires, which are furnished either solid or stranded, are normally used for heavy-power circuits or as service-entrance leads to buildings. Wire sizes larger than these are used for extremely heavy loads and for poleline distributions. Shown are the allowable current-carrying capacity for copper and aluminum conductors. Shown, too, is the percent reduction in current capacity if more than three conductors are in a cable or raceway.

Multiconductor Cables. There are many cases where the use of individual conductors, spaced and supported side by side, becomes an inefficient and hazardous practice. For these installations, multiconductor cables consist of the individual conductors as described above, arranged in groups of two or more. An additional insulating or protective shield is formed or wound around the group of conductors. The individual conductors are color coded for proper identification. The illustrations provided show some of the types of multiconductors.

ARMORED CABLE, commonly referred to as BX, can be supplied either in two or three wire types and with or without a lead sheath. The wires in BX, matched with a bare equipment ground wire, are initially twisted together. This grouping, totaling three or four wires when combined with the ground, is then wrapped in coated paper and a formed self-locking steel armor. The cable without a lead sheath is widely used for interior wiring under dry conditions. The lead sheath is required for installation in wet locations, and through masonry or concrete building partitions for added protection for the copper conductor wires.

NONMETALLIC SHEATHED CABLE consists of two or three rubber or thermoplastic-insulated wires, each covered with a jute type of filler material which acts as a protective insulation against mishandling. This in turn is covered with an

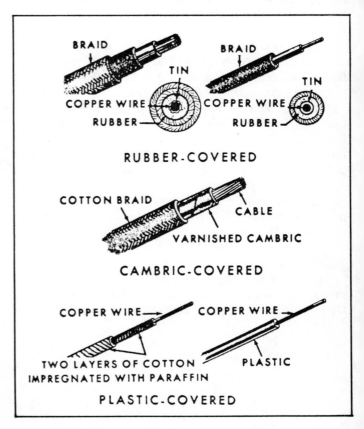

Single conductors

Interior Wiring

impregnated cotton braid. The cable is lightweight, simple to install, and comparatively low priced. It is used quite extensively in interior wiring, but is not approved for use in wet locations. A dual-purpose plastic sheathed cable with solid copper conductors can be used underground, outdoors or indoors. It needs no conduit, and its flat shape and gray or ivory color make it ideal for surface wiring. It resists moisture, acid, and corrosion and can be run through masonry or between studding.

LEAD-COVERED CABLE consists of two or more rubber-covered conductors surrounded by a lead sheathing which has been extruded around it to permit its installation in wet and underground locations. Lead-covered cable can also be immersed in water or installed in areas where the presence of liquid or gaseous vapors would attack the insulation on other types of cable.

PARKWAY CABLE provides its own protection from mechanical injury and therefore can be used for underground services by burying it in the ground without any protecting conduit. It normally consists of rubber-insulated conductors enclosed in a lead sheath, and covered with a double spiral of galvanized steel tape which acts as a mechanical protection for the lead. On top to the tape, a heavy braid of jute saturated

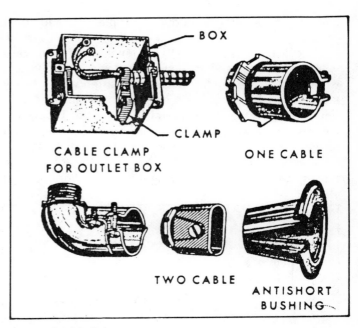

Armored-cable fittings

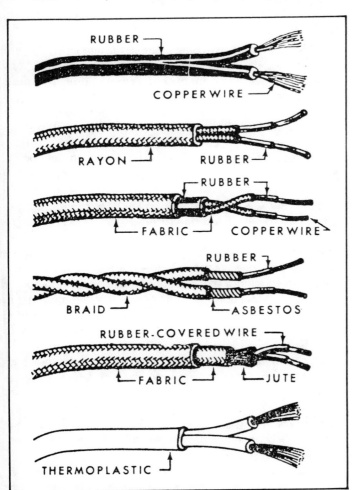

Types of flexible cords

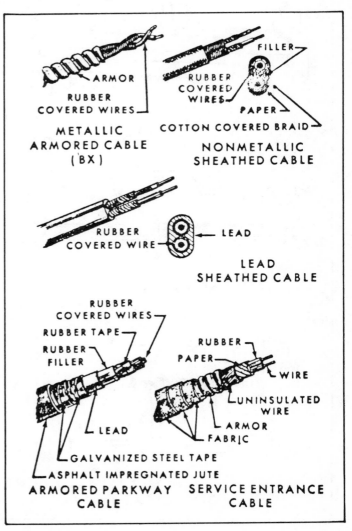

Multiconductor cables

SECTION II: Setting Up the Basic Farm Structure

with a waterproofing compound is applied for additional weather protection.

SERVICE-ENTRANCE CABLE normally has three wires with two insulated and braided conductors laid parallel and wound with a bare conductor. The wires are encased in heavy tape or armor for protection; the tape serves as an inner cushion, and covers the whole assembly with braid. Though the cable normally serves as a power carrier from the exterior service drop to the service equipment of a building, it may also be used in interior-wiring circuits to supply power to electric ranges and water heaters at voltages not exceeding 150 volts to ground provided the outer covering is armor. It may also be used as a feeder to other buildings on the same premises under the same conditions, if the bare conductor is used as an equipment grounding conductor from a main distribution center located near the main service switch.

Cords. Many items using electrical power are either of the pendant, portable, or vibration type. In these cases the use of cords as shown is authorized for delivery of power. These can be grouped and designated as either lamp, heater, or heavy-duty power cords. Lamp cords are supplied in many forms. The most common types are the single-paired rubber-insulated and twisted-paired cords. The twisted-paired cords are of two cotton-wound conductors which have been covered with rubber and rewound with cotton braid. Heater cords are similar to this latter type except that the first winding is replaced by heat-resistant asbestos. Heavy-duty or hard-service cords are normally supplied with two or more conductors surrounded by cotton and rubber insulation. During manufacture, these are first twisted or stranded. The voids created in the twisting process are then filled with jute and the whole assembly covered with rubber. All cords, whether of this type or of the heater or lamp variety, have the conductors color coded for easy identification. Shown by common trade terms are the cords found in general use.

Electrical Boxes

Design. Outlet boxes bind together the elements of a conduit or armored cable system into a continuous grounded system. They provide a means of holding the conduit in position, a space for mounting devices such as switches and receptacles, protection for the device, and space for making splices and connections. Outlet boxes are manufactured in sheet steel, porcelain, bakelite, or cast iron and can be round, square, octagonal or rectangular. The fabricated steel box is available in a number of different designs. For example, some boxes are of the sectional or "gang" variety, while others have integral brackets for mounting on studs and joists. Certain boxes have been designed to receive special cover plates so that switches, receptacles, or lighting fixtures may be more easily installed. Other designs ease installation in plastered surfaces. Regardless of the design or material, they all should have enough interior space to allow splicing of conductors or making of connections. For this reason the minimum depth allowed 1½ in. in all cases, except where building-supporting members would have to be cut to leave this much room. In this case the minimum depth can be reduced to ½ in.

Attachment Devices for Outlet Boxes. Outlet boxes which do not have brackets are supported by wooden cleats or bar hangers as shown.

Wooden cleats. Wooden cleats are first cut to size and nailed between two wooden members. The boxes are then either nailed or screwed to these cleats through holes provided in their back plates.

Strap hangers. If the outlet box is to be mounted between studs, mounting straps are necessary. The ready-made straps are handy and accommodate not only a single box, but a 2-,

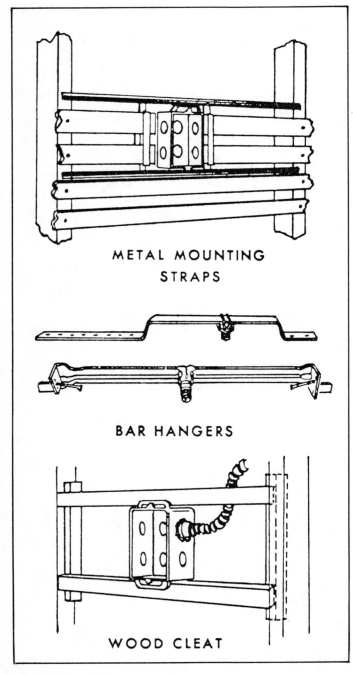

Typical box mountings

Interior Wiring

3-, 4-, or 5-gang box.

Bar hangers. Bar hangers are prefabricated to span the normal 16-in. and 24-in. joist and stud spacings and are available for surface or recessed box installation. They are nailed to the joist or stud-exposed faces. The supports for recessed boxes are called offset bar hangers.

Patented supports. When boxes have to be installed in walls that are already plastered, any one of several patented supports can be used for mounting. These eliminate the need for installing the boxes on wooden members, and thus avoid extensive chipping and replastering.

Knobs, Tubes, Cleats, Loom and Special Connectors. Open wiring requires the use of special insulating supports and tubing to insure a safe installation. These supports, called knobs and cleats, are smooth-surfaced and made of porcelain. Knobs and cleats support the wires which are run singly or in pairs on the surface of the joists or studs in the buildings. Tubing or tubes, as they are called, protect the wires from abrasion when passing through wooden members. Insulation of loom of the "slip-on" type is used to cover the wires on box entry and at wire-crossover points. The term "loom" is applied to a continuous flexible tube woven of cambric material impregnated with varnish. At points where the type of wiring may change and where boxes are not specifically required, special open wiring to cable or conduit wiring connectors should be used. These connectors are threaded on one side to facilitate connection to a conduit, and have holes on the other side to accommodate wire splices but are designed only to carry the wire to the next junction box. (Specific methods of installation were covered earlier.)

Cable and Wire Connectors. Code requirements state that "Conductors shall be spliced or joined with splicing devices approved for the use or by brazing, welding, or soldering with a fusible metal or alloy. Soldered splices shall first be so spliced as to be mechanically and electrically secure without solder and then soldered." Soldering or splicing devices are used as added protection because of ease of wiring with and high quality of connection of these devices. Assurance of high quality is the responsibility of the electrician who selects the proper size of connectors relative to the number and size of wires. Illustrations show some types of cable and wire connectors in use.

Straps and Staples.
Policy. All conduits and cables must be attached to the structural members of a building in a manner that will prevent sagging. The cables must be supported at least every 4½ ft. for either a vertical or horizontal run and must have a support in the form of a strap or staple within 12 in. of every outlet box. Conduit-support spacings vary with the size and rigidity of the conduit.

Cable Staples. A very simple and effective method of supporting BX cables on wooden members is by use of cable staples, as shown.

Selection. The selection of boxes in an electrical system should be made in accordance with tables which list the

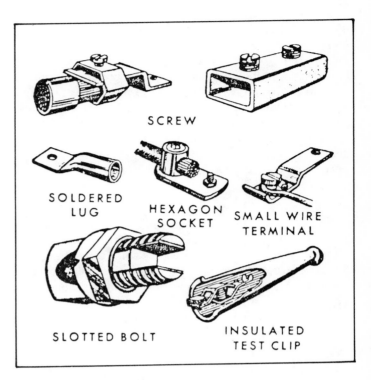

Cable and wire connectors

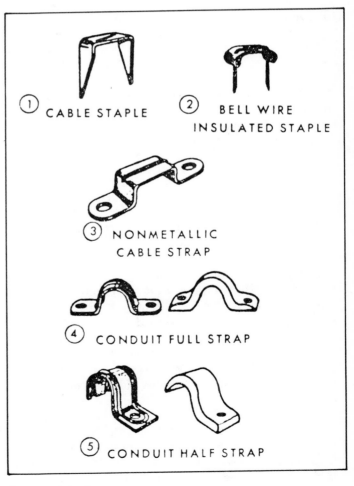

Straps and staples

maximum allowable conductor capacity for each type of box. As shown in these illustrations, each conductor running through the box is counted along with each conductor terminating in the box. For example, one conductor running through a box and two terminating in the box would equal three conductors in the box. Consequently, you could choose any of the boxes listed. This applies to boxes that do not contain receptables, switches, or similar devices. Each of these items mounted in a box reduces by one the maximum number of conductors allowed.

Outlet Boxes for Rigid and Thin-Wall Circuit and Armored Cable. Steel or cast iron outlet boxes are generally used with rigid and thin-wall conduit or armored cable. The steel boxes are either zinc- or enamel-coated, the zinc coating being preferred when installing conduit in wet locations. All steel boxes have "knockouts." These knockouts are indentations in the side, top, or back of an outlet box, sized to fit the standard diameters of conduit fittings or cable connectors. They usually can be removed with a small cold chisel or punch for easier entry into the box of the conduit or cable. Boxes designed specifically for armored-cable use also have integral screw clamps located in the space immediately inside the knockouts, and thus eliminate the need for cable connectors. This reduces the cost and labor installation. Box covers are normally required when it is necessary to reduce the box openings, provide mounting lugs for electrical devices, or to cover the box when it is to be used as a junction. Illustrations show several types of cable connectors and also a cable clamp for use in clamping armored cable in an outlet box. The antishort bushing shown in the illustrations is inserted between the wires and the armor to protect the wire from the sharp edges of the cut armor when it is cut with a hacksaw or cable cutter.

Outlet Boxes for Nonmetallic Sheathed Cable and Open Wiring Steel. Steel boxes are also used for nonmetallic cable and open wiring. However, the methods of box entry are different from those for conduit and armored-cable wiring because the electrical conductor wires are not protected by a hard surface. The connectors and interior box clamps used in nonmetallic and open wiring are formed to provide a smooth surface for securing the cable, rather than being the sharp-edged type of closure normally used.

Nonmetallic. Nonmetallic outlet boxes are of either porcelain or bakelite may also be used with open or nonmetallic sheathed wiring. Cable or wiring entry is generally made by removing the knockouts of preformed weakened blanks in the boxes.

Note. In open wiring, conductors should normally be installed in a loom from the last support to the outlet box. Although all of the boxes described in the above are permissible for open wiring, a special loom box is available which has its back corners "sliced off" and allows for loom and wire entry at this sliced-off position.

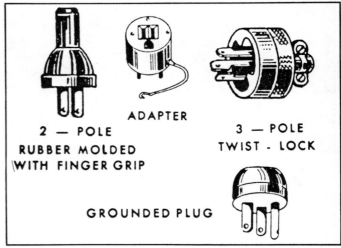

Attachment plugs

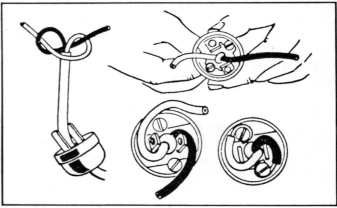

The Underwriters knot, used for plugs, prevents the wires from pulling off of the leads and through the plug. When pulled back into the plug, the knot prevents it being pulled back through. First, draw cord through the plug and tie knot as at upper left. Trim sufficient insulation from the wire to conveniently allow for each lead being attached to a screw, with the wire circling clockwise around the screw; tighten.

Plugs and Cord Connectors. Plugs. Portable appliances and devices that are to be connected to receptacles have their electrical cords equipped with plugs that have prongs which mate the slots in the outlet receptacles. A three-prong plug can fit into a two-prong receptacle by using an adapter. If the electrical conductors connected to the outlet have a ground system, the lug on the lead wire of the adapter is connected to the center screw holding the receptacle cover to the box. Many of these plugs are permanently molded to the attached cords. There are other types of cord-grips that hold the cord firmly to the plug. Twist-lock plugs have prongs that catch and are firmly held to a mating receptacle when the plugs are inserted into the receptacle slots and twisted. Where the plugs do not have cord-grips, the cords should be tied with an Underwriters knot at plug entry; this eliminates tension on the terminal connections when the cord is connected and disconnected from the outlet receptacle. Shown are the steps to be used in tying this type of knot.

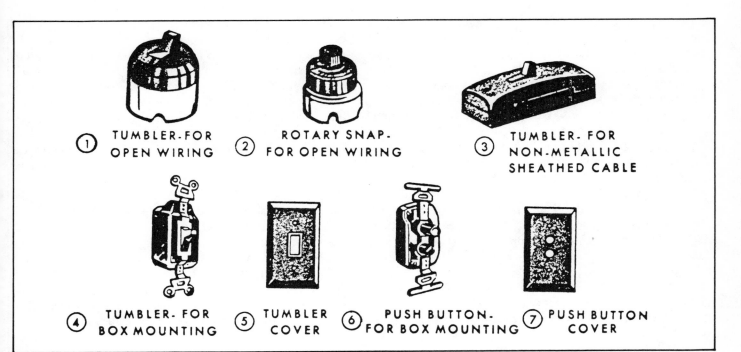

Switches and covers

Cord Connectors. There are some operating conditions where a cord must be connected to a portable receptable. This type of receptacle, called a cord connector body or a female plug, is attached to the cord in a manner similar to the attachment of the male plug outlined above.

Switches and Covers. Definition. A switch is a device used to connect and disconnect an electrical circuit from the source of power. Switches may be either one-pole or two-pole for ordinary lighting or receptacle circuits. If they are one-pole, they must be connected to break the hot or ungrounded conductor of the circuit. If two-pole, the hot and ground connection can be connected to either pole on the line side of the switch. Switches are also available that can be operated in combinations of two, three, or more in one circuit. These are called three-way and four-way switches.

Open and Nonmetallic Sheathed Wiring. Switches used for exposed open wiring and nonmetallic sheathed cable wiring are usually of the tumbler type with the switch and cover in one piece. Other less common ones are the rotary-snap and push-button types. These switches are generally nonmetallic in composition.

Conduit and Cable Installations. The tumbler switch and cover plates normally used for outlet-box installation are mounted in a manner similar to that for box-type receptacles and covers, and come in two pieces.

Entrance Installations. At every powerline entry to a building there is a switch-and-fuse combination or circuit-breaker switch installed at the service entrance. This switch must be operated to disconnect the building load while in use at the system voltage. Entrance or service switches, as they are commonly called, consist of one "knife" switch blade for every hot wire of the power supplied. The switch is generally

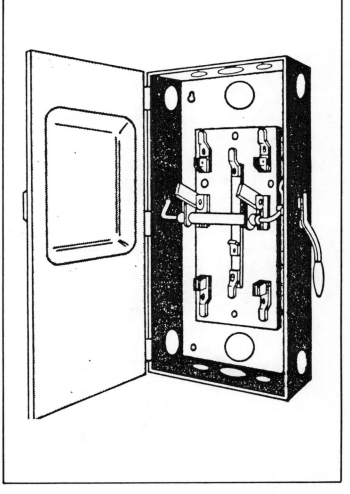

Service switch box

enclosed and sealed in a sheet-steel cabinet. When connecting or disconnecting the building circuit, the blades are operated simultaneously through an exterior handle by the rotation of a common shaft holding the blades. The neutral or grounded conductor is not switched, but is connected as a neutral terminal within the box. Many entrance switches are equipped with integral fuse blocks or circuit breakers which protect the building load. The circuit breaker entrance switch is preferred, particularly in field installations, because it is easy to reset after the overload condition in the circuit has been cleared.

Fuses and Fuse Boxes. Fuses. The device for automatically opening a circuit when the current rises beyond the safety limit is technically called a cutout, but more commonly called a fuse. All circuits and electrical apparatus must be protected from short circuits or dangerous overcurrent conditions through correctly rated fuses.

STANDARD. The cartridge fuse is used for current rating above 30 amperes in interior wiring systems. The ordinary plug or screw fuse is satisfactory for incandescent lighting or heating appliance circuits.

SPECIAL. On branch circuits, wherever motors are connected, time-lag fuses should be used instead of the standard plug or cartridge fuse. These fuses have self-compensating elements which maintain and hold the circuit in line during a momentary heavy ampere drain, yet cut out the circuit under normal short-circuit conditions. The heavy ampere demand normally occurs in motor circuits when the motor is started. Examples of such circuits are the ones used to power oil burners or air conditioners.

FUSE BOXES. As a general rule the fusing of circuits is concentrated at centrally located fusing or distribution panels. These panels are usually at the service-entrance switch in small buildings, or installed in several power centers in large buildings. The number of service centers or fuse boxes in the latter case would be determined by the connected power load.

Circuit Breaker Panels. Circuit breakers are devices resembling switches that trip or cut out the circuit in case of overamperage. They perform the same function as fuses, and can be obtained with time-lag opening features similar to the special fuses mentioned above. They may be classified as thermal or magnetic reaction types.

A thermal type circuit breaker has a bimetallic element integrally built within the breaker that responds only to fluctuations in temperature within the circuit. The element is made by bonding together two strips of dissimilar metal, each of which has a different rate of expansion. When a current is flowing in the circuit, the heat created by the resistance of the bimetallic element will expand each metal at a different rate, causing the strip to bend. The element acts as a guard in the circuit; the breaker mechanism is adjusted so that the element bends just far enough under a specified current to trip the breaker and open the circuit.

A magnetic circuit breaker responds to changes in the magnitude of current flow. In operation an increased current flow will create enough magnetic force to "pull up" an armature, opening the circuit. The magnetic circuit breaker is usually used in motor circuits for closer adjustments to motor rating, while the circuit conductors are protected by another circuit breaker.

There is also a thermal-magnetic breaker, which as the name implies, combines the features of the thermal and magnetic types. Practically all of the molded case circuit breakers used in lighting panelboards are of this type. The

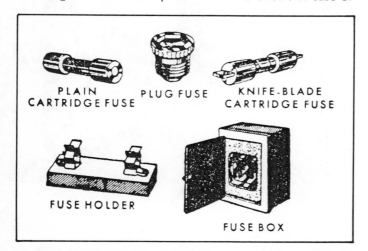

Typical fuses and fuse box

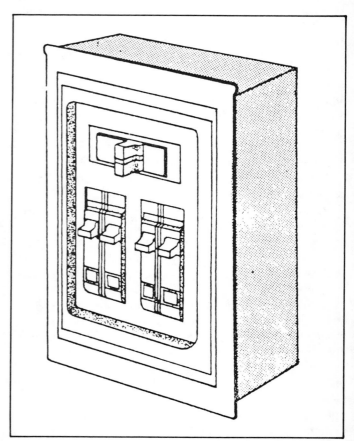

Typical circuit breaker box

thermal element protects against overcurrents in the lower range while the magnetic element protects against the higher range, usually resulting from short circuits. During the last decade, circuit breakers have been used to a greater extent than fuses because they can be manually reset after tripping, whereas fuses require replacement. Also may easily be replaced with higher capacity ones that do not protect the circuit. This is difficult to do with circuit breakers. In addition they combine the functions of fuse and switch, and when tripped by overloads or short circuits, all of the ungrounded conductors of a circuit are opened simultaneously.

Each branch circuit must have a fuse or circuit breaker protecting each ungrounded conductor. Some installations may or may not have a main breaker that disconnects everything. As a guide during installation, if it does not require more than six movements of the hand to open all the branch circuits breakers, a main breaker or switch is not required ahead of the branch-circuit breaker. However, if more than six movements of the hand are required, a separate disconnecting main circuit breaker is required placed ahead of the branch-circuit breaker. Each 120-volt circuit requires a single-pole (one-pole) breaker which has its own handle. Each 208-volt circuit requires a double-pole (two-pole) breaker to protect both ungrounded conductors. You can, however, place two single-pole breakers side by side, and tie the two handles together mechanically to give double-pole protection. Both handles can then be moved by a single movement of the hand. A two-pole breaker may have one handle or two handles which are mechanically tied together, but either one requires only one movement of the hand to break the circuit.

Insulating Staples. Bell or signal wires are normally installed in pairs in signal systems. The operating voltage and energy potential is so low in these installations (12 to 24 volts) that protective coverings such as conduit or loom are not required. To avoid any possibility of shorting in the circuit, they are normally supported on wood joists or studs by insulated staples.

Straps. Conduit and cable straps are supplied as either one-hole or two-hole supports and are formed to fit the contour of the specific material for which they are designed. The conduit and cable straps are attached to building materials by "anchors" designed to suit each type of supporting material. For example, a strap is attached to a wood stud or joist by nails or screws. Expanding anchors are used for box or strap attachment to cement or brick structures and also to plaster or plaster-substitute surfaces. Toggle and "molly" bolts are used if the surface wall is thin and has a concealed air space which will allow for the release of the toggle or expanding sleeve.

Receptacles, Fixtures, and Receptacle Covers.

Applicability. Portable appliances and devices are readily connected to an electrical supply circuit by means of an outlet called a receptacle. For interior wiring these outlets are installed either as single or duplex receptacles. Receptacles previously installed and their replacements in the same box, may be two-wire receptables. All others must be the three-wire type. The third wire on the three-wire receptable is used to provide a ground lead to the equipment which receives power from the receptacle. This guards against dangers from current leakage due to faulty insulation or exposed wiring, and helps prevent accidental shock. The receptacles are constructed to receive plug prongs either by a straight push action or by a twist-and-turn push action. Fixtures are similar to receptacles but are used to connect the electrical supply circuit directly to lamps inserted in their sockets.

Knob-and-Tube Wiring. Receptacles that have their entire enclosures made of some insulating material, such as bakelite, may be used without metal outlet boxes for exposed, open wiring or nonmetallic sheathed cable.

Conduit and Cable. The receptacles commonly used with conduit and cable installations are constructed with yokes to aid their installation in outlet boxes. In this case they are attached to the boxes by metal screws through the yokes, threaded into the box. Wire connections are made at the receptacle terminals by screws which are an integral part of the outlet. Receptacle covers made of either brass, steel or nonmetallic materials are then attached to box and receptacle installations to afford complete closure at the outlets.

Surface Metal Raceways. These provide a quick, inexpensive electrical wiring installation method since they are installed on the wall surface instead of the inside the wall.

1. Surface metal raceway is basically of two types: one-

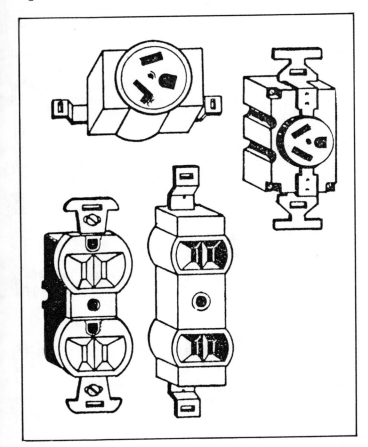

Types of wall receptacles

SECTION II: Setting Up the Basic Farm Structure

stalled like conduit, then the wires are "pulled" to make the necessary electrical connections. If working with the two-piece construction type, the base piece is installed along the wiring run. Wiring is next laid in the base piece and held in place with clamps. After the wires are laid, the capping is snapped on and the job is complete.

2. A multi-outlet system, with grounding inserts if desired, has outlets spaced every few inches so that several tools or pieces of equipment can be used simultaneously. An over-floor metal raceway system handles telephone and signal or power and light wiring where the circuits must be brought to locations in the middle of the floor area. These systems are designed so that they can be installed independent of other wiring systems, or economically connected to existing systems.

Transformers. The transformer is a device for changing alternating current voltages into either high voltages for efficient powerline transmission or low voltages for consumption in lamps, electrical devices, and machines. Transformers vary in size according to their power handling rating. Their selection is determined by input and output voltage and load current requirements. For example, the transformer used to furnish power for a doorbell reduces 115-volt circuit and two secondary screw terminals from the low voltage side of the transformer. It is used to lower the building voltage of 120 volts or 240 volts ac to the 6, 12, 18, or 24 volts ac. The wires shown here are input and output leads. Also shown are the input leads which are smaller than the output leads because the current in the output circuit is greater than in the input circuit.

Lighting

Lampholders and Sockets

Lamp sockets shown are generally screw-base units placed in circuits as holders for incandescent lamps. A special type of lampholder has contacts, rather than a screw base, which engage and hold the prongs of fluorescent lamps when they are rotated in the holder. For open wiring installations, the sockets are attached to a hanging cord or mounted directly on a wall or ceiling by using screws or nails in the mounting holes found in the nonconducting material molded or formed around the lamp socket. The two mounting holes in a porcelain lamp socket are spaced so the sockets may also be attached to outlet box "ears," or a plaster ring with machine screws. The screw threads molded or rolled in the ends of the lampholder sockers also aid their use in other types of lighting fixtures such as table lamps, floor lamps, or hanging fixtures which have reflectors or decorative shades. In an emergency, a socket may also be used as a receptacle. The socket is converted to a receptacle by screwing in a female plug. One type of ceiling lampholder has a grounded outlet located on the side.

Lamp sockets are produced in many different sizes and shapes. Their normal operating voltages are 6, 12, 18, or 24 volts, ac or dc. As a general rule they are connected by open-wiring methods and are used as interoffice or building-to-building signal systems.

Reflectors and Shades

Shown are several types of reflectors and shades used to focus the lighting effect of bulbs. Of these, some are used to flood an area with high-intensity light and are called floodlights. Others, called spotlights, concentrate the useful light on small areas. Both floodlights and spotlights can come in two- or three-light clusters with a swivel holders. They can be mounted on walls or posts or on spikes pushed into the ground.

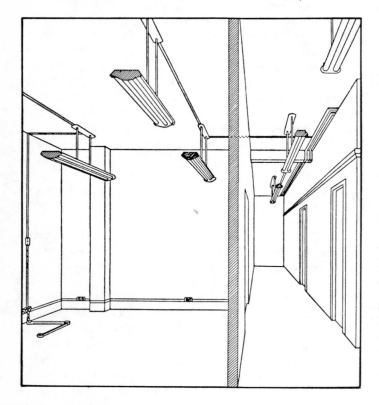

Surface metal raceways

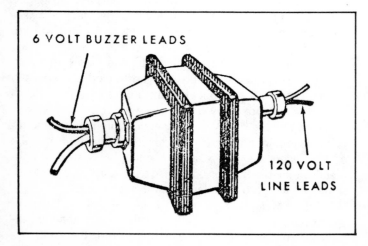

Transformer

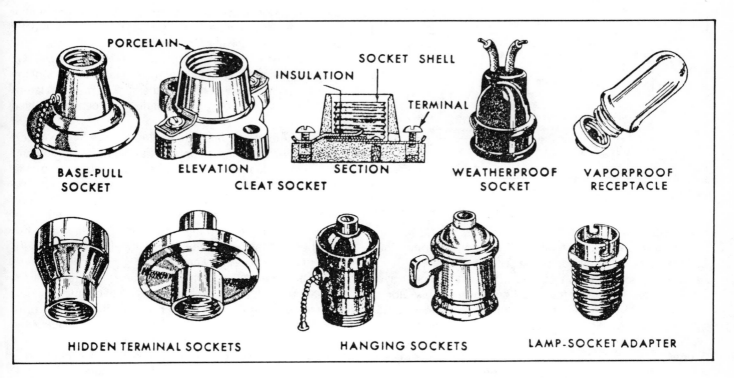

Lampholders and sockets

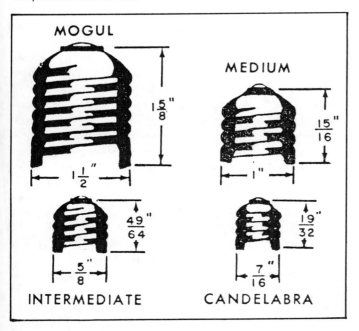

General lamp-socket sizes

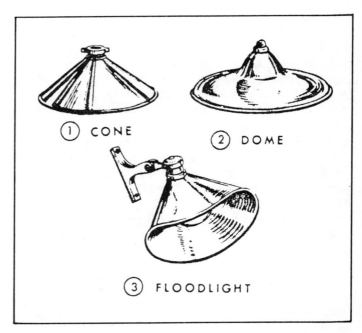

Types of reflectors

Incandescent Lamps

The most common light source for general use is the incandescent lamp. Though it is the least efficient type of light, it is preferred because of its low initial cost, ease of maintenance, convenience, and flexibility. Its flexibility and convenience are seen by the wide selection of wattage ratings that can be inserted in one type socket. Further, since its emitted candle-power is directly proportional to the voltage, a lower voltage application will dim the light. A high-rated voltage application from a power source will increase its intensity. Although an incandescent light is economical, it is also inefficient because a large amount of the energy supplied to it is converted to heat rather than light. Moreover, it does not give a true light because the tungsten filament emits a great deal more red and yellow light than does the light of the sum. Incandescent lights are normally built to last 1,000 hours when operating at their rated voltage.

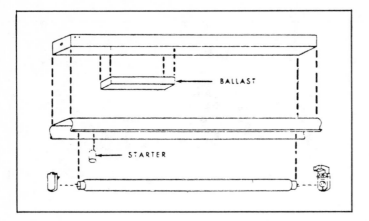

Fluorescent Lamps

Fluorescent lamps are either of the conventional "hot cathode" or "cold cathode" type. The "hot cathode" lamp has a coiled wire type of electrode, which when heated gives off electrons. These electrons collide with mercury atoms, provided by mercury vapor in the tube which produces ultraviolet radiation. Fluorescent powder coatings on the inner walls of the tubes absorb this radiation and transform the energy into visible light. The "cold cathode" lamp operates in a similar manner, except that its electrode consists of a cylindrical tube. The heat is generated over a larger area and, therefore, the cathode does not reach as high a temperature as in the "hot cathode" tube. The "cold cathode" is less efficient but has a longer operating life than the "hot cathode" unit. It is used most frequently on flashing circuits. Because of the higher light output per watt input, more illumination and less heat is obtained per watt from fluorescent lamps than from incandescent ones. Light diffusion is also better and surface brightness is lower. The life of fluorescent lamps is also longer compared to filament types. However, the fluorescent lamp, because of its design, cannot control its beam of light as well as the incandescent type and tends to flicker. Moreover, when voltage fluctuations are severe, the lamps may go out prematurely or start slowly. Finally, the higher initial cost in fluorescent lighting, which requires auxiliary equipment such as starters, ballasts, special lamp holders and fixtures is also a disadvantage.

1. Detail illumination is required where the intensity of general illumination is not sufficient. The fixtures for detail illumination commonly use single fluorescent lamps. Illustrated are the wiring arrangements for these single units, and a multiple unit.

2. Because of greater cost and shorter life of 8-watt fluorescent lamps, as compared to 15-watt and 20-watt lamps, fixtures with 8-watt lamps are used only for detail illumination and general illumination within locations where space is restricted.

3. Although the fluorescent lamp is basically an ac lamp, it can be operated on dc with the proper auxiliary equipment. The current is controlled by an external resistance in series with the lamp. Since there is no voltage peak, starting is more difficult and thermal switch starters are required. The lamp tends to deteriorate at one end due to the uniform direction of the current. This may be partially overcome by periodically reversing the lamp position or the direction of current.

4. Because of the power lost in the resistance ballast box in the dc system, the overall lumens per watt efficiency of the dc system is about 60 percent of the ac system. Also, lamps operated on dc may only give as little as 80 percent of rated life.

5. The fluorescent lamp, like all discharge light sources, requires special auxiliary control equipment for starting and stabilizing the lamp. This equipment consists of an iron-core choke coil, or ballast, and an automatic starting switch connected in series with the lamp filaments. The starter (starting switch) can be either a glow switch or a thermal switch. A resistor must be connected in series with the ballast in dc circuits, because the ballast alone does not offer sufficient resistance to maintain a steady arc current.

6. Each lamp must be provided with an individual ballast and starting switch, but the auxiliaries for two lamps are usually enclosed in a single container. The auxiliaries for fluorescent lighting fixtures are mounted inside the fixture above the reflector. The starting switches (starters) project through the reflector so that they can be replaced easily. The circuit diagram for the fixture appears on the ballast container.

Characteristics. A hot-cathode fluorescent lamp gradually loses electron-emissive material from the electrodes; this loss is accelerated by frequent starting. The rated average life of the lamp is based on normal burning periods of 3 to 4 hours.

EFFICIENCY. The efficiency of a fluorescent lamp is very sensitive to changes in temperature of the bulb. The maximum efficiency occurs in the range of 100° F. to 120° F., which is the operating temperature that corresponds to an ambient room temperature range of 65° F. to 85° F. The efficiency decreases slowly as the temperature rises above normal, but also decreases very rapidly as the temperature falls below normal. Thus, the fluorescent lamp should not be located where it will be subjected to wide variations in temperature. Reduced efficiency due to lower room temperature can be minimized by operating the fluorescent lamp in a tubular glass enclosure.

Fluorescent lamps are relatively efficient compared with incandescent lamps. For example, a 40-watt fluorescent lamp produces approximately 2800 lumens, or 70 lumens per watt. A 40-watt fluorescent lamp produces six times as much light per watt as does the comparable incandescent lamp.

Fluorescent lamps should be operated at voltage within 8 percent of their rated voltage. If the lamps are operated at lower voltages, uncertain starting may result. If operated at higher voltages, the ballast may overheat. Operation of the lamps at either lower or higher voltages results in decreased lamp life.

When lamps are operated on ac circuits, the light output undergoes cyclic pulsations; this produces a flicker that may cause a "strobe" effect. This can be minimized by combining two or three lamps in a fixture and operating them on different phases of a three-phase system.

SECTION 3: Specialized Needs

12. Building Greenhouses and Propagating Frame

Almost every gardener eventually reaches a point where he wants a greenhouse. Before buying or building one, you should:
(1) give careful thought ot the size, style, and kind of climate controls desired;
(2) contact your county agricultural agent so he can help you locate and visit some of your neighbors with garden greenhouses to learn about their problems;
(3) check local building codes and zoning laws.

A greenhouse can range from a simple polyethylene covered framework that you can put together in an afternoon for less than fifty dollars, to a six-thousand dollar fully automated conservatory. No matter which size or type of greenhouse you choose, consider how much time you'll have to spend in it after it's built. Don't be over enthusiastic; some new greenhouse owners find they do not have as much time as they thought for gardening. On the other hand, there is a misconception that greenhouses require constant attention. By combining automatic controls and easy-care plants, maintenance can be kept to an hour a week. Automatic controls are ideal for providing proper growing temperature, artificial light, watering, humidity, and ventilation. Or, if you have the time, you can save a lot of money by not using automatic controls.

You can get the most greenhouse for your money by doing some of the construction work yourself. How much work you do depends on how handy you are with tools. Be honest with yourself—don't take on a job that's too big to handle. If you are good with tools, you can put up any plastic-covered greenhouse, and almost any prefabricated glass greenhouse. You should, however, hire a qualified plumber and electrician.

Selection and Sites

There are two basic types of greenhouses: attached and free standing. An attached greenhouse may be even-span, lean-to, or window mounted. A freestanding greenhouse is usually even-span (symmetrical roof).

Attached Lean-to
A lean-to greenhouse is built against a building, using the existing structure for one or more of its sides. It is usually attached to a house, but may be attached to other buildings.

The usual means of attachment is a ledge of thick lumber to which roof rafters can be attached.

The lean-to is limited to single or double-row plant benches with a total width of 7 to 12 ft. It can be as long as the building to which it is attached. The advantage of the lean-to greenhouse is that is usually is close to available electricity, water and heat.

Disadvantages are that the lean-to offers only limited space, limited light, limited ventilation, and limited temperature control.

Attached Even-Span
The even-span greenhouse is the type people generally visualize when they think of a greenhouse. The even-span greenhouse is similar to a free-standing structure except that it is attached to a house at one gable end. It can accommodate two or three rows of plant benches.

The cost of an even-span greenhouse is greater than the cost of a lean-to, but it has greater flexibility in design and provides for more plants. Because of its increased size and exposed glass area, the even-span greenhouse will cost more to heat.

Attached Window-Mounted
A window-mounted greenhouse will allow space to grow a few plants at a relatively low cost in heating and cooling. This reach-in greenhouse is available in many standard sizes, either in single units or in tandem arrangements for large windows. Only simple tools are needed to remove the regular window from the frame and to fasten the prefabricated window greenhouse in its place.

Free-Standing
The free-standing greenhouse is a separate structure and consists of side walls, end walls, and gable roof. It is like an even-span except that a free-standing greenhouse is set apart from other buildings in order to receive more light. It can be built large or small.

A separate heating system is necessary unless the greenhouse is very close to a heated building. The free-standing greenhouse is more easily adapted to the builder's desired location, size and shape than attached greenhouses. It also utilizes more sunlight, but requires more heat at night due to the additional glass.

SECTION III: Specialized Needs

According to a laboratory report, this energy-saving Vegetable Factory lean-to greenhouse can produce tomatoes at a cost of only 6¢ a pound. (Art courtesy of Pat Wagner Communication Services.)

Lean-to Structures

If you decide to construct a lean-to addition, a good portion of yours is already taken care of since you are working with an existing structure.

The lean-to is really nothing more than half of a freestanding structure, attached to another building. If it is a greenhouse, (a popular lean-to addition) be sure the house siding is of a treated material that will resist decay and deterioration due to humidity, and that you have kept dangerous pesticides used in the greenhouse from entering the home.

The lean-to uses one house wall, and is three-sided. It can be of any design, but should match the character of the house. Use a structural element of the main house, such as a post or moldings, to tie the two together design-wise.

The lean-to may require a door that leads to the main house; these can be aluminum siding doors, or wooden. The floor can be concrete or brick.

Most important will be the joint where the roof of the lean-to attaches to the supporting building. Remove the siding on the existing house, insert metal flashing to avoid leaks, then replace the siding. The roof can be a continuation of the house roof, with double 2x10 rafters bolted to the pillars of concrete, or 4x4s. If of concrete, they can be 8x8s. Be sure the floor is prepared according to the standard for floors mentioned elsewhere in the book.

As mentioned, one common addition is a lean-to greenhouse; We therefore have included here artwork for adding a lean-to greenhouse. For a shed, or even an open stall, the building basics would be the same and the finishing aspects would be different.

When building a lean-to, the important thing to keep in mind is that you are simply building a half of a framing structure, but that you will be essentially following the same types of steps as in building a complete structure when it comes to procedures for foundations, walls and roof.

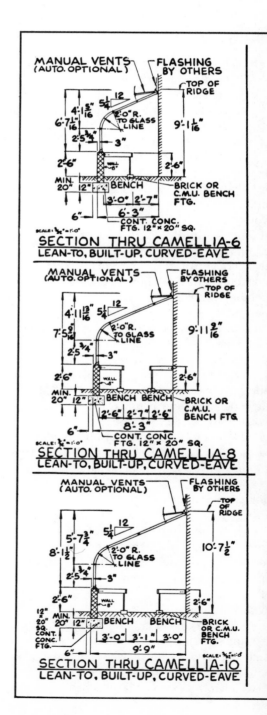

Building Greenhouses

A greenhouse attached to the main house is the most economical type to construct, and often the most convenient because it provides easier access from the home. If adding a greenhouse, take steps to protect your home from incipient humidity.

SECTION III: Specialized Needs

Even-span greenhouse available through J.A. Nearing Co. (Janco).

Locating Your Greenhouse

The first choice for a greenhouse site should be on the south or southeast side of your house, in a sunny location. The east side is the second best location. That's where it will capture the most November-to-February sunlight. The next best locations are the southwest and west. The north side is the least desirable location. Some plants will grow in a greenhouse no matter where it has been located. African violets and orchids will grow with northern exposure, but heating costs will be high. You limit the types of plants you can grow by not putting the greenhouse in the best possible location.

You can place your greenhouse where it will be partly shaded during the summer when light reduction is not serious and may even be desirable. Be sure to take into account the possibility of falling tree limbs that can damage the greenhouse.

Sometimes you can place a greenhouse against a door, window, or basement entrance of your house. This will let you use heat from your house to grow plants, make your greenhouse more accessible, and save on construction costs. Your home heating bill, however, will increase significantly.

If you have an L-shaped house, you can save the cost of two greenhouse walls by building the greenhouse in the "L".

Whether your greenhouse runs north/south or east/west is not as important as wind protection. Protect your greenhouse from winds by locating it so existing buildings will shield it, or by providing it with a windbreak hedge or fence.

An ideal site for your greenhouse would be one that is well-drained, nearly level, and has full exposure to sunlight. It would slope slightly to the south and have a windbreak on the side of the prevailing wind.

Height

The height of the greenhouse depends on the desired height to the eave. An eave height of 5 ft. is satisfactory for side benches with low-growing plants. If you want to grow tall plants, however, you will want an eave height of 6 or 7 ft.

The pitch of the roof should be 6 in 12 (approximately 27 degrees). The eave height, the distance from the side wall to the center of the greenhouse, and the roof pitch will determine the height of your greenhouse at the center. The height of the greenhouse should be equal to eave height plus one-fourth the width of the greenhouse. For instance, in an even-span greenhouse 18 ft. wide, the distance from the side wall to the center of the greenhouse is 9 ft. The difference in height between the center of the greenhouse and the eave will be one-half of 9 ft., or 4 ft. If the eave is 5 ft. high, the greenhouse should be 9½ ft. at the center.

Cost

The lowest cost per square foot of growing space is available with an even-span greenhouse 17 to 18 ft. wide. It will house two side benches, two walks, and a wide center bench.

The lowest total-cost greenhouse is the lean-to house 7 to 12 ft. wide with double-row benches and a central walk.

Designing Your Greenhouse

Width

Width is the most important dimension; it will not change during the life of the greenhouse. Length can be increased if more space is desired.

Determine the width of your greenhouse by adding the widths of the plant benches and the walks. Allow approximately 6 in. for walls at either side and 2 in. for an air-circulation space between the side walls and the benches.

Side benches are serviced from only one side and should be no wider than you can reach across. For some people this will be 2 ft., for others perhaps as much as 3 ft. Center benches are serviced from both sides and can be as wide as 6 ft. They should be no wider than you can work comfortably.

Determine the width of the walks in your greenhouse by how they are to be used. If the walks will be used only as a place to stand while servicing the benches, an 18- or 19- in. walk is sufficiently wide; if a wheelbarrow will be brought into the greenhouse, the width must be greater. Wide walks—24 to 30 in.—will allow easy passage for visitors who may not be used to walking between rows of plants.

Length

Determine the length of your greenhouse by mulitplying the number of plants you can grow across the benches by the number of plants you want to grow. Then round off the measurement so that no glass will need to be cut to fill odd sash bar spacings. (A sash bar is a shaped wooden or metal bar used in the construction of a sash or frame and designed to hold and support the glass secure to it.)

Standard glass sizes are 16x24, 18x20, and 20x20 in. (Larger glass sizes means fewer sash bars and less shadow inside the greenhouse.) Most plastics are available in 100-ft. lengths.

When you figure the length of a glass greenhouse, allow for the width of the projecting part of each sash bar, plus a fraction of an inch clearance. For plastic, allow an extra 24 in. to fasten the plastic properly.

Greenhouse Construction Basics

Whether you build a glass, fiberglass, or plastic greenhouse, it will pay you to shop around.

Greenhouses have supporting framework made of wood, aluminum, iron, or galvanized pipe. Some have curved eaves; others have flat eaves. Some are glass or plastic from the ground up. All types have advantages and disadvantages.

If you build your own greenhouse, have the plumbing and electrical work done by professionals in accordance with local codes. Most local governments require a building permit to erect a greenhouse.

Glass Greenhouse

Glass is the traditional greenhouse covering. It is available in many designs and blends with almost any architecture. Glass greenhouses may have slanted sides, straight sides and eaves, or curved eaves.

Aluminum, maintenance-free glass construction has very pleasing lines and will provide a large growing area. It assures you of a weathertight structure, which minimizes heat costs and retains humidity.

For amateur gardeners, small prefabricated glass greenhouses are available for do-it-yourself installation. They are sold in different models, to fit available space and to fit your pocketbook.

The disadvantages of glass are that it is easily broken, expensive, and requires more care in construction than fiberglass or plastic.

Fiberglass Greenhouses

Fiberglass is lightweight, strong, and practically hailproof. Corrugated panels 8 to 12 ft. long and flat fiberglass in rolls are available in 24- to 48-in. widths. Thicknesses range from 3/64 to 3/32 of an inch.

Poor grades of fiberglass will discolor and the discoloring reduces light penetration. Using a good grade, on the other hand, may make your fiberglass greenhouse as expensive to build as a glass one. If you select fiberglass, choose the clearest grade. Do not use colored fiberglass.

Plastic Greenhouses

Plastic greenhouses are increasing in popularity. This is because construction cost per square foot is generally one-sixth to one-tenth the cost of glass greenhouses, and plastic

SECTION III: Specialized Needs

Plastic-covered greenhouse layout

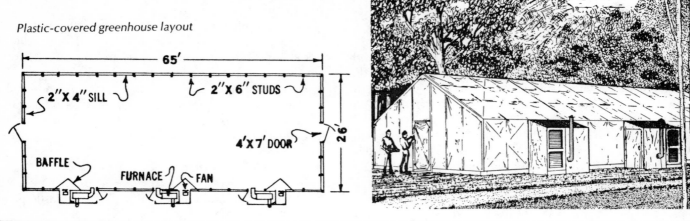

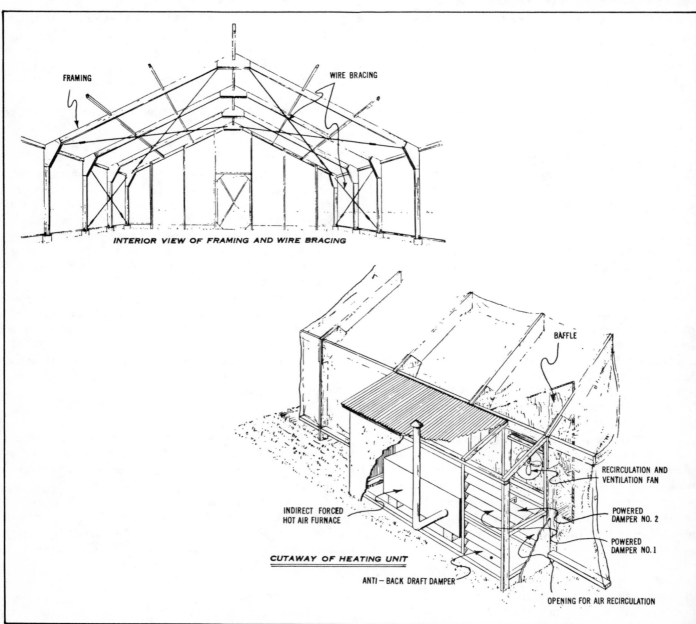

Plastic-covered greenhouse construction details

greenhouses can be heated as satisfactorily as glass greenhouses. Also, crops grown under plastic are of equal quality to those grown under glass, even though plastic greenhouses are considered temporary structures and usually carry a low assessment rate for tax purposes, or may not be taxed at all.

Plastic greenhouses can be made of polyethylene (PE), polyvinyl chloride (PVC), copolymers of these materials, and other readily available clear films. Polyethylene must be replaced each year; it deteriorates rapidly in the strong sunlight of summer. Other films such as PVC or copolymers with ultraviolet (UV) inhibitors last longer. Descriptions of plastics available for covering hobby greenhouses are provided below.

Polyethylene. Polyethylene is low in cost and lightweight. It also stands up well in fall, winter and spring weather, and lets through plenty of light for good plant growth. However, polyethylene constantly exposed to the sun deteriorates during the summer and must be replaced each year. Ultraviolet light energy causes polyethylene to break down. This first deterioration occurs along (or over) the rafters and along the creases where the film is folded.

Ultraviolet-inhibited polyethylene lasts longer than regular polyethylene. It has an inhibitor that prevents the rapid breakdown caused by ultraviolet light. UV-inhibited polyethylene is available in 2- and 6-mil thicknesses up to 40 ft. wide and 100 ft. long.

Polyethylene permits passage of much of the reradiated heat energy given off by soil and plants inside the greenhouse. Therefore a polyethylene greenhouse loses heat more quickly than a glass greenhouse both during sunny periods and after sunset. This is an advantage during the day but a disadvantage at night.

Polyvinyl chloride (PVC or Vinyl). Vinyls fron 3 to 12 mils thick are available for greenhouse covering. Like polyethylene, vinyls are soft and pliable; some are transparent, others transluscent. Vinyl attracts dust and dirt from the air and has to be washed occasionally. They are usually available in 4 to 6 ft. widths only; larger widths can be made by electronically sealing several smaller widths together.

Vinyls cost from two to five times as much as polyethylene. When carefully installed, 8- or 12-mil vinyl holds up for as long as 5 years.

Frames for Plastic Greenhouses

Plastic greenhouse structures range from crude wooden frameworks to air-supported houses. If you plan to build a plastic greenhouse, carefully consider economy of size and future expansion.

Because plastic is available in large widths and is lighter in weight, greenhouse rafters and supporting members can be widely spaced to permit maximum light penetration. Common types of greenhouse frames are as follows.

A-Frame. In building an A-frame structure, attention must be given to the placement of cross rafters (supporting members). Cross rafters should be placed at least one-third of the distance down from the ridge on the outer rafters. Otherwise, it will be difficult to work around the cross rafters when applying an insulating layer of plastic.

When the cross-rafter support is high in the peak of the greenhouse—especially in narrow greenhouses—an essentially clear-span type of structure permits easy application of an inner layer of plastic. The inner layer can be applied under the cross-rafter supports, leaving a small triangular air space in the peak of the house.

Diagonal bracing wires add strength to an A-frame structure, which is among the least difficult to build.

Rigid Frame. Rigid-frame structures have been designed in widths up to 40 feet. This clear span design has no columns to hold up the roof section. The best available rigid-frame greenhouse has a 6-ft. sidewall and is designed for 30, 36, or 40-ft. widths.

A prefabricated greenhouse built with curved laminated wood rafters is commercially available. It has very low sidewalls (low head room), and to grow tall plants the structure must be raised higher on the foundation sidewalls.

Panel Frame. Panel-frame greenhouses are a modification of the sash house (a small plastic greenhouse used for growing plants for later transplanting). This structure requires accurate carpentry, and construction costs are higher than for other frames because of the added lumber and labor needed to build the panels.

Advantages of panels are that they can be quickly installed and taken down and stored during the summer; this will increase the life of the plastic panels. Panel greenhouses can be easily ventilated.

Quonset. Quonset greenhouses have the same general shape as the quonset huts of World War II. Some have been constructed of wood, but usually the frames are metal. The half-circle frames are covered with one piece of wide plastic and the houses are up to 20 ft. wide. The advantage of this house is its ease of construction and covering. Ventilation is by exhuast fans at the ends of the house.

Pipe Frame. A pipe structure can be used to frame an air-inflated greenhouse. Air is introduced into a chamber formed by two layers of 4- or 6-mil film.

The effect of the air under slight pressure is to force the inner layer of film over the circular greenhouse pipe frames. The outer layer assumes a circular shape over the frame and rides on a cushion of air. It lifts 3 to 4 in. from the frame at the top and 1 to 2 in. from the frame at the foundation sill. Air enters the chamber through 6-in. plastic tubing. A manometer is used to measure static air pressure between the two layers of film.

Weatherproof Wire

Use weatherproof wire for all outside wiring. Wire size depends upon the distance to be covered and the number of hotbeds to a circuit. Use approved terminal equipment and follow safe wiring practices. All wiring must conform to local wiring codes.

SECTION III: Specialized Needs

Environment Control

Heating

You must decide which of the many heating systems available best suits your greenhouse operation. Consider the initial cost, economy of operation, and available fuel. You can heat your greenhouse efficiently with coal, electricity, gas, and oil. Heating equipment can be a space heater, a forced-air heater, a hot-water or steam system, or electric heaters. Radiant heat lamps over plants and soil-heating cable under plants can also be used.

The capacity of your heating system will depend on the size of your greenhouse, whether it is covered with a single layer or a double layer of plastic or glass, and the maximum difference between inside and outside temperatures.

Heating systems are rated in British thermal units (B.t.u.'s) per hour. The firm from which you buy your greenhouse can tell you what size of heater you will need, or you can follow these steps to estimate the size.

First, find the temperature difference. This is the difference in degrees Fahrenheit between the lowest outside temperature and the temperature you want to maintain inside your greenhouse. For instance, if you want to maintain a minimum inside temperature of 60 degrees and the coldest night temperature you expect is −10 degrees, your temperature difference is 70 degrees.

Next, find the number of square feet of exposed glass or plastic in your greenhouse. Don't forget to add the areas of the sides and ends to the area of the roof. Then multiply the temperature difference by the number of square feet. For example, suppose you have a 20-by-100-ft. greenhouse with a total of 3,400 sq. ft. of exposed plastic. You would multiply 3,400 by 70 (the temperature difference). This would give you 238,000.

Now, if your greenhouse is covered with two layers of plastic or glass, multiply the 238,000 by 0.8. If it is covered with only one layer, multiply by 1.2. This will give the required B.t.u. per hour capacity of your heater. In the example, a two-layer greenhouse would be 238,000 x 0.8 equals 190,400 B.t.u.'s per hour. The one layer greenhouse would be 238,000 x 1.2 equals 285,600 B.t.u.'s per hour.

The type of heating system your choose will depend on how much you want to spend. There are four types.

Space heaters. For low-cost heating of small greenhouses, use one or more ordinary space heaters. WARNING: If you use a gas, oil, or coal heater, be sure to have a fresh air supply to avoid carbon monoxide buildup due to restricted oxygen supply. Fans are also needed to improve circulation. Use high grade (low sulfur) kerosene to avoid sulfur dioxide damage; the need for high ignition temperature to avoid carbon monoxide and ethylene buildup is important.

Forced-air heater. The best system for heating a small greenhouse is a forced-air furnace with a duct or plastic tube system to distribute heat. You can use a thermostat to control the temperature in the greenhouse.

Hot-water or steam heater. A hot-water system with circulator or a steam system linked with automatic ventilation will give adequate temperature control. In some areas, coal or natural gas is readily available at low cost. This fuel is ideal for hot-water or a central steam system. Steam has an advantage in that it can be used to sterilize growing beds and potting soils.

Electric heaters. Overhead infrared heating equipment combined with soil cable heat create a localized plant environment which allows plants to thrive even though the surrounding air is at a lower-than-normal temperature. Electric resistance-type heaters are used as space heaters or in a forced-air system.

Ventilation

Even during cold weather a greenhouse can get too warm on bright, sunny days. So ventilation equipment should be built into your greenhouse to control temperatures in all seasons. If you use hand-operated roof vents, they will require frequent temperature checks. As outdoor weather changes, sashes must be opened and closed manually to keep plants from getting too hot or cold.

Automatic ventilation eliminates the manual work and is the best way to cool a greenhouse. As an example, if your greenhouse has roof vents a special electric motor and thermostat will open and close the vents. Fresh outside air is brought in through the roof vents. Warm air flows out through escape vents. Besides cooling the greenhouse the change of air improves growing conditions. Responding to this air transfer, the thermostat will turn off and on to keep temperatures right for plants.

Fans provide good ventilation and are needed in both large and small greenhouses. Exhaust fans should be large enough to change the air in the greenhouse once every minute. To accomplish this, the capacity of the fan in cubic feet per minute at 1/8-in. static pressure should equal the volume of the greenhouse. The volume can be calculated by multiplying the floor area by 7.

If the greenhouse is high enough, place the exhaust fan and the motorized intake louvers above the doors at opposite endwalls. This will exhaust the hottest, most humid air, and prevent a direct draft on the plants near the intake.

Fan and duct ventilation can also be used for automatic greenhouse heating and ventilation. Plastic ducts are suspended by wires or straps from the roof of the greenhouse. The fan-heater-louver unit gives positive air flow and the polyethylene duct distributes the incoming air evenly throughout the house.

Shade

When protection from the sun is needed, use rollup screens of wood or aluminum, vinyl plastic shading, or paint-on materials. Roll-up screens are available with pulleys and rot-resistant nylon ropes. These screens can be easily adjustable from outside, as weather and sunlight vary.

Vinyl plastic shading is made of a flexible film that reduces light from 55 to 65 per cent. The material comes in rolls and installs easily against the glass inside your greenhouse. To apply, just wash the glass with a wet sponge, then smooth the plastic onto the wet glass. When smoothed into position it adheres to the glass. It can be pulled off and used again.

Shading compound can also be applied on the outside of glass greenhouses. It can be thinned with paint solvents. It usually comes in choices of white or green. Shading compound which mixes with water can also be used.

Evaporative Cooling

An evaporative cooler (or fan and pad system) eliminates excessive heat and adds beneficial humidity to the greenhouse atmosphere. With an evaporative cooler, moist cool air is circulated throughout the greenhouse. Warm air flows out through roof vents or exhaust fans. Temperature is lowered, humidity is increased, and watering needs are reduced.

To select a cooler of the right size:

(1) calculate the cubic feet of your greenhouse by multiplying the length by the width by the average height;
(2) add 50 percent to the total cubic space, then select a cooler which has at least this CFM (cubic feet per minute) air capacity rating.

The cooler must be installed outside the greenhouse. If it is inside, it can only humidify and cannot cool. A properly sized cooler will reduce the greenhouse temperature approximately 80 percent of the difference between the outside wet-bulb and dry-bulb thermometer readings.

In hot, dry areas this system can reduce the temperature from 30 to 40 degrees. In wet, humid areas the cooling will be less. It is most effective during the hottest part of the day.

Mist Propagation Controls

Mist sprays are used in propagating to keep the atmosphere humid. There are two types of mist propagation controls. The most popular is by means of time clocks. The other system controls the cycles by evaporation from a mechanical or electronic leaf or screen.

Time clock system. This system of automatic watering includes: a dual-time clock consisting of a 24-hour clock and a 6-minute clock; an electric water valve with strainer; hose bibbs; a toggle switch for a choice of manual or automatic operations.

Evaporation System. This system provides a special unit that operates within the mist spray from the nozzles. When the stainless steel or ceramic screen and the plants become saturated, the screen tilts to a downward position, which switches off the water.

The water evaporates both on the mesh screen and on the cuttings. When the screen loses weight, the screen raises and actuates the switch. This opens the solenoid valve and starts the misting cycle again, according to the needs of the cuttings. Because this control is activated by the weight of the water, it is fully automated and operates continuously day and night.

Watering kits for pot plants. Watering kits for pot plants can also be used. Water is supplied directly to each plant through hollow plastic tubes, which are permanently attached. One tube can be used for each small pot and two or more for larger pots. Water tubes are weighted at the outlet end; each tube is approximately 5 ft. long, and can be cut to shorter length if necessary.

CO_2 Control

Carbon dioxide (CO_2) and light are needed for plant growth. Closed greenhouses often have too little carbon dioxide during the day to effectively utilize available light. Therefore, plants grow poorly when air vents are closed.

By enriching the atmosphere with CO_2, plant growth can be accelerated.

Because light and carbon dioxide complement each other in plant growth, additional electric lights in greenhouses combined with good carbon dioxide control will increase yields of lettuce, tomatoes, orchids, chrysanthemums, carnations, snapdragons, geraniums, and other crops.

CO_2 equipment utilizing infrared sensors are available for greenhouse owners who want to benefit from carbon dioxide enrichment with supplementary lighting. The equipment will measure and control CO_2 levels from 0 to 2,000 parts per million, which will satisfy most of the production needs of greenhouse growers. This equipment is fairly expensive and requires frequent calibration.

Inexpensive color metric kits are also available for determining the CO_2 levels in your greenhouse.

Forms of CO_2 for enriching greenhouse atmospheres can be purchased in the forms listed here.

Bottled CO_2. Liquefied from a burning process, this CO_2 is kept under pressure and is controlled by means of a solenoid or metering device.

Dry ice. This form is convenient because it may be placed in a greenhouse or growth chamber in block form, or placed in a pressure bottle and stored until needed.

Burned sulfur-free gaseous fuels. These include natural gas, LP gas, or a liquid carbon fuel such as kerosene.

Light Control

Plants respond to the relative lengths of light and dark periods as well as to the intensity and quality of light. Artificial light has been used extensively to control plant growth processes under various conditions.

Plants differ in the need for light; some thrive on sunshine, others grow best in the shade. Most plants will grow in either natural or artificial light.

Artificial light in greenhouses can be used to provide high-intensity light when increased plant growth is desired, or to extend the hours of natural daylight, or to provide a night interruption to maintain the plants on long-day conditions.

Proper lighting not only extends the gardening day by enabling the gardener to work in the greenhouse during the dark evenings of winter and early spring, but it aids plant growth. Three basic types of lamps are used in greenhouse lighting.

Fluorescent. The most widely used lamp for supplemental light, it offers higher light efficiency with low heat. Available in a variety of colors, cool-white lamps are the most commonly used. High-intensity (1500 ma) fluorescent tubes that require higher wattage are also commonly used to reach 2000 foot candles.

Incandescent. Used to extend day length in greenhouses, the lamps range from 60 to 500 watts. The grower can vary footcandle levels by adjusting the spacing and mounting height above the plants.

High-intensity discharge (HID). These have a long life (5000 hours or more). With improvements made possible by the addition of sodium and metal halides, the lamp has a high emission of light in the regions utilized by plants.

Light Meters. Inexpensive light meters are available for measuring the light intensity in greenhouses. The most common light meters are calibrated in foot candles or lux (10.76 foot candles).

Temperature

As a gardener you will be concerned with two temperatures; the air temperature required in the greenhouse, and the minimum outside temperature that your heating equipment must overcome.

For most plants, a night temperature of 60 to 65° F. in the greenhouse is adequate. The general rule, however, is not to have a higher temperature than is necessary.

If you grow some plants that require a higher temperature than is provided in the greenhouse, use a propagating case or a warmed bench with thermostatically controlled warming cables to warm the air surrounding the plants. This can be done at a fraction of the cost that would be necessary to heat the whole greenhouse to provide the same temperature. Space heaters can maintain a minimum of 60° F. in the greenhouse. Higher temperatures on plant benches can be provided with soil-warming equipment.

If you want a temperature of 60° F., install heaters that will provide that temperature. If you want no more than frost protection, set the thermostat at 40° F.

Remember that heat is lost from a greenhouse by radiation and conduction, plus convection through: glass; walls and other non-glass parts of the structure; floor or soil; ventilation, door openings, and cracks.

Control Units

Automatic controls are important in greenhouses. Without them switching lights, fans, pumps, heaters, and misters on and off at prescribed times would be a complicated and laborious task.

Many time clocks, photocells, thermostats, and other controls are available commercially. When used individually or in combination they will provide any time interval or control desired.

Project: Plastic-Covered Greenhouse

This plan is applicable to most areas in the south-eastern United States. Its use in other areas may require some modification of the structure and of the environmental control devices.

The building is of wooden, rigid-frame construction, with glued and nailed plywood gussets. The rigid frames are made of 2 x 4 studs, and are securely anchored to circular, concrete piers with U-type steel straps.

The wooden 2 x 4 sill extends between studs, but the 2 x 4 plate is continuous throughout the length of the greenhouse.

The structure, intended to be covered with plastic film, has been designed in 4-ft. increments so it can also accommodate rigid plastic panels if the builder desires.

In a high-moisture environment, the wooden frames of greenhouses are subject to rapid decay. Therefore, treat wood with a nontoxic preservative.

Beds for Growing Small Plants

Coldframes

A coldframe is a bottomless box with a removable top. It is used to protect small plants from wind and low temperatures. No artificial heat or manure is used inside a coldframe.

Coldframes utilize the sun's heat. The soil inside the box is heated during the day and gives off its heat at night to keep the plants warm. The frame may be banked with straw or strawy manure to insulate it from the outside air and to retain heat.

With a coldframe, you can do many of the same things you do in a greenhouse. You can sow summer flowers and vegetables weeks before outdoor planting. Often, you will gain sufficient time to grow an extra crop. You can start vegetables, annual flowers for fall and winter, and perennials for next year's bloom. Plants are protected from harsh weather and will grow to transplant size quickly. You can root cuttings of deciduous and evergreen shrubs and trees. Softwood cuttings of chrysanthemums, geraniums, and fuchsia, and leaf cuttings of rex begonias. African violets, and succulent and foliage plants take root faster in a coldframe, particularly, during warmer months. You can also grow your own lettuce, chives, endives, parsley, and green onions right through the winter by converting your coldframe to a hotbed.

Portable coldframes can be built in your workshop from surplus materials you may have on hand. Coldframes are constructed from sections of 3- by 4-ft. or 3- by 6-ft. millwork sash or plastic-covered panels. Most coldframes can be converted to hotbeds for use in all seasons by installing electric heat, and automatic clock-controlled misting or watering.

Hotbeds

A hotbed is a bed of soil enclosed in a glass or plastic

Hotbed and Propagating Frame

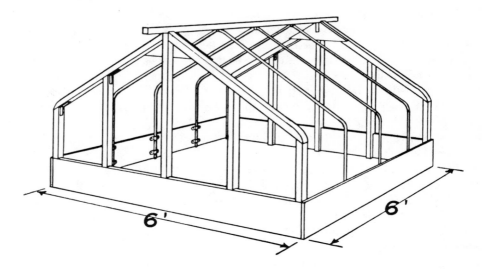

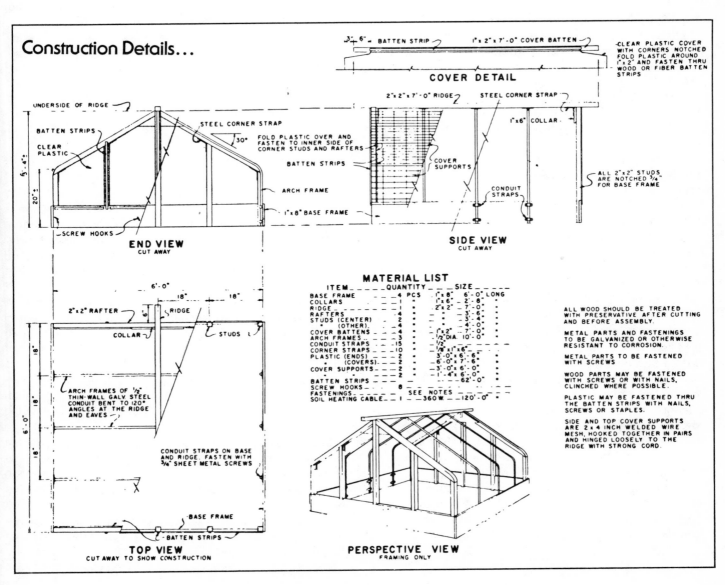

frame. It is heated by manure, electricity, steam or hot-water pipes.

Hotbeds are used for forcing plants or for raising early seedlings. Instead of relying on outside sources of supply for seedlings, you can grow vegetables and flowers best suited to your own garden. Seeds may be started in a heated bed weeks or months before they can be sown out of doors. At the proper time the hotbeds can be converted into a coldframe for hardening. Then the plants may be moved to the garden when outdoor conditions are favorable.

Between 10 and 15 watts of electric heat should be provided for every square foot of growing area in a hotbed. Soil-heating tape or cable is available in several lengths, with a choice of wattages. Tape or wire screening, ¼- or ½-in. mesh, should be placed over the heating tape or cable to prevent possible damage from cultivating tools.

If the bed is in a sunny, well-sheltered location, and the climate not too severe, 10 watts per square foot should be adequate. For localities with very cold winters, a higher heat capacity is needed; 15 watts per square foot is recommended. Line side walls with moisture-proof insulation.

Do not place hotbed cables of any type directly in peat. When peat dries out it acts as an insulator and may cause the cable to overheat. Use a thermostat to control temperature automatically and make more efficient use of electricity.

Because accurate temperature control is possible with a thermostat, you can grow better plants at lower costs by separating plants requiring different temperatures in different beds.

Temperatures from 50 to 70° F. are best for hotbeds. On very cold nights cover the beds with mats, burlap, straw, or boards.

Construction of Hotbed and Propagating Frame

This plastic-covered hotbed, which can also be used as a propagating frame, is inexpensive and easy to build.

The hotbed has welded wire frames with removable covers of 4-mil polyethylene plastic film on each side of the ridge. The covers may be rolled down from the ridge or up from the sides to provide almost unlimited adjustment for ventilation. The covers are secured in the desired position with light ropes that pass over the sides and down to the base at each end of the frame. Rubber tubing can be attached at one end of each rope to provide tension. When the spring season is over, the covers can be rolled up and stored out of sunlight to prevent deterioration from ultra-violet radiation. The plastic film that covers the end sections of the structure is not removable, so a long-lasting plastic such as 4-mil vinyl or 3-mil type W polyester film is recommended. Details for making the removable covers are given the the working drawings.

The hotbed is made of wood. It has three arch frames of thin-wall electrical conduit. The frames may be easily bent to shape with a hand conduit bender. The wooden parts should be treated with a preservative after they have been cut. A thorough soaking in a 5-percent solution of pentachlorophenol is suggested.

A soaking trough for treating the wood can be improvised by lining a small trench with polyethylene film. Weight the boards to be sure thay are completely immersed in the liquid. Place thin spacers between the boards so the preservative will flow completely around each peice.

PRECAUTION: Keep children and animals away from the soaking trough. Follow directions given on the label for use of preservative.

The hotbed should be located on well-drained soil. Some locations may require a 3-in. layer of gravel under the prepared soil mixture of flats. For supplementary heating, try a 360 watt electric soil-heating cable with a 70° F. thermostat. Lay the cable on a bed of sand or vermiculite and cover with about 2 in. of sand. If the seed or cuttings are to be started in a prepared soilbed rather than in pots or flats, protect the cable by placing ½-in.-mesh hardware cloth about 1 in. above it.

The propagating frame. When the hotbed is used as a propagating frame, the welded wire frame is covered with cheesecloth fastened with clothespins. Two mist sprayers, fastened to a 2 x 2 board 8 ft. long, can be mounted quickly and easily on the inside of the frame, diagonally and about 1 ft. above the cuttings, or on the outside above the ridge. The location of the sprayers depends on the weather. If it is dry and windy, the sprayers are mounted inside. If humid and calm, the sprayers are mounted on the ridge. The frame is kept covered with cheesecloth, and the cheesecloth is moistened several times a day when necessary.

Project: Simple Greenhouse

This easy-to-construct greenhouse measures 8 x 12 ft. It features two potting shelves 2 ft. wide and 12 ft. long, three ceiling cross-braces designed to support hanging plants, and 96 sq. ft. of floor space to store larger plants and shrubs, planting supplies, and gardening equipment. Two screen doors and adjustable ceiling vents assure proper ventilation and good control of humidity ranges.

To build this greenhouse you need just seven panels of plywood, lumber framing, plastic sheeting, and the basic woodworking skills.

Building Hints

These general hints are designed to help you achieve the best possible results in working with plywood. They apply not only to this plan, but to all projects you may undertake using plywood.

Layout. Following the panel layout, draw all parts on the plywood panels using a straightedge and a carpenter's square for accuracy. Use a compass to draw corner radii. Be sure to allow for saw kerfs when plotting dimensions; if in doubt, check the width of your saw cut.

Cutting. For hand-sawing use a 10 to 15 pt. cross-cut. Support panel firmly with face up. Use a fine-toothed coping saw for curves. For inside cuts, start hole with drill and then use a coping or keyhole saw. For power sawing, a plywood

Simple Greenhouse

blade gives best results; however, a combination blade may be used. Panel face down for hand power sawing. Panel face up for table power sawing. With first cuts reduce panel to pieces small enough for easy handling. Use scrap lumber beneath the panel (clamped or tacked securely in place) to prevent splintering on back side. Plan to cut matching parts with same saw setting. If available, you may use a jigsaw, bandsaw, or sabre saw for curved cuts. In any case be sure blade enters face of panel.

Drilling. Support plywood firmly. For larger holes use brace and bit. When point appears through plywood, reverse and complete hole from back. Finish slowly to avoid splintering.

Planing. Edge grain of plywood runs in alternate directions, so plane from ends toward center. Use shallow-set blade.

Sanding. Most sanding should be confined to edges with 1-0 or finer sandpaper, before sealer or flat undercoat is applied. You may find it easier to sand the cut edges smooth before assembling each unit. Plywood is sanded smooth in manufacture—one of the big time-savers in its use—so only minimum surface sanding is necessary. Use 3-0 sandpaper in direction of grain only, after sealing.

Assembly. Assemble by sections. For example, drawers, cabinet sheels, compartments, or any part that can be handled as an individual completed unit. Construction by section makes final assembly easier. For strongest possible joints, use a combination of glue and nails (or screws); to glue-nail, check for a good fit by holding the pieces together. Pieces should contact at all points for lasting strength. Mark nail locations along edge of piece to be nailed. In careful work, where nails must be very close to an edge, you may wish to predrill using a drill bit slightly smaller than nail size. Always predrill for screws. Apply glue to clean surfaces, according to manufacturer's instructions. Press surfaces firmly together until "bead" appears, then nail, check for square, and apply clamps if possible to maintain pressure until glue sets. For exterior exposure, use resorcinol-type waterproof glue; for interior work, use liquid resin (white) or urea resin type glues. (Other glues are available for special problems.)

SECTION III: Specialized Needs

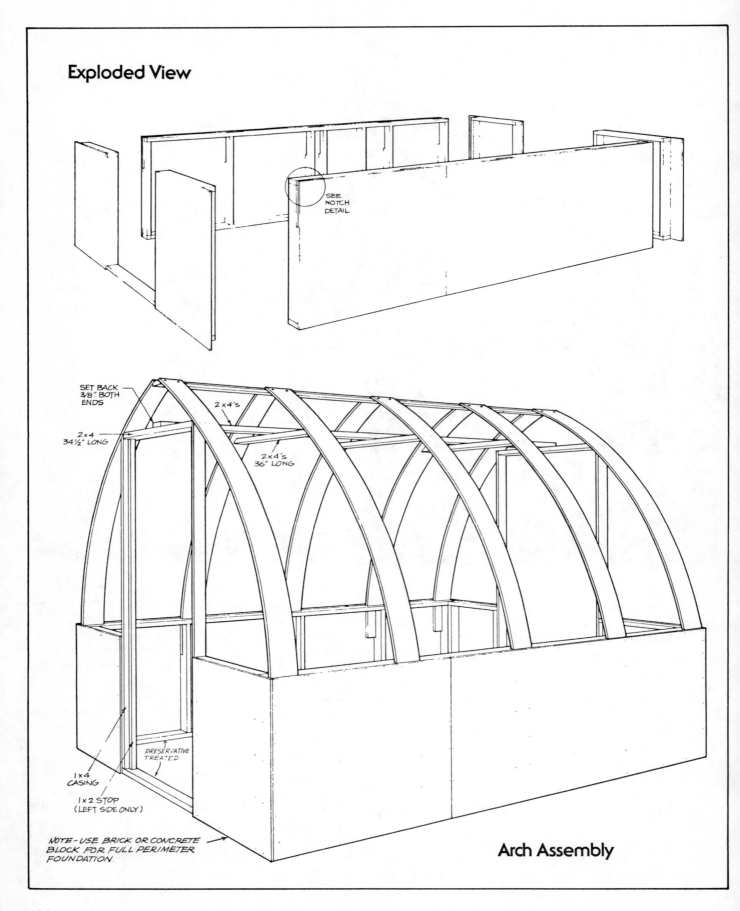

Building Greenhouses

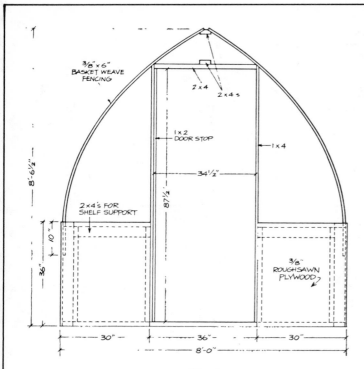

Ends

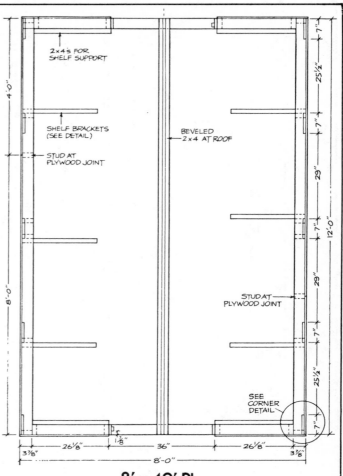

8' x 12' Plan

Materials List

Recommended plywood: 303⁽ⁱ⁾ textured plywood siding with desired surface texture, APA grade-trademarked.
A-C or C-C Exterior plywood, APA grade-trademarked for shelves.

PLYWOOD

Quantity	Description
5 panels	3/8 in. x 4 ft x 8 ft Exterior plywood siding
2 panels	1/2 in. x 4 ft x 8 ft A-C or C-C Exterior plywood

OTHER MATERIALS

Quantity	Description
80 lin. ft	10, 3/8 in. x 6 in. x 8 ft basket weave-type fencing
228 lin. ft	2 x 4 lumber for framing
24 lin. ft	1 x 2 lumber stiffening for shelves
30 lin. ft	1 x 4 lumber for door casing
68 lin. ft	2 x 2 lumber for door framing
12 lin. ft	1 x 2 lumber for door insert framing
15 lin. ft	1 x 2 lumber for door stop framing
As required	Plastic for seasonal use: 6 to 10 mil polyethylene. One continuous piece 12 ft wide x 17 ft long. Two pieces 8 ft x 8 ft for doors and ends.
Optional	For more permanent installation use corrugated fiberglass panels applied horizontally on the roof and down to the sides — applied with corrugations running vertically on the ends and doors.
2 pieces	Approximate size: 36 x 42 in. sections of screening
6	Door hinges (3 per door) 3-1/2" butt hinges
4	Vent hinges (2 per vent) 1-1/2" butt hinges
10	2 in. galvanized carriage bolts
40	1 in. round-head machine screws (washers not necessary)
As required	1 in. construction nails
As required	Caulk, sealant or flashing material to fill any joints, etc.
As required or desired	Finishing materials - stain or paint, wood preservative (see Building Hints)

NOTE: In areas of heavy snow, do not allow buildup on the roof.

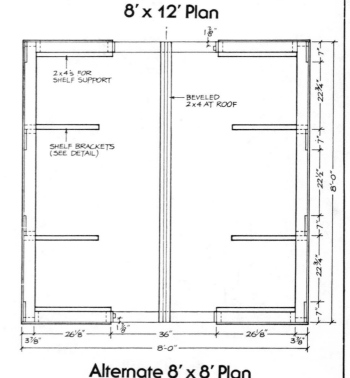

Alternate 8' x 8' Plan

SECTION III: Specialized Needs

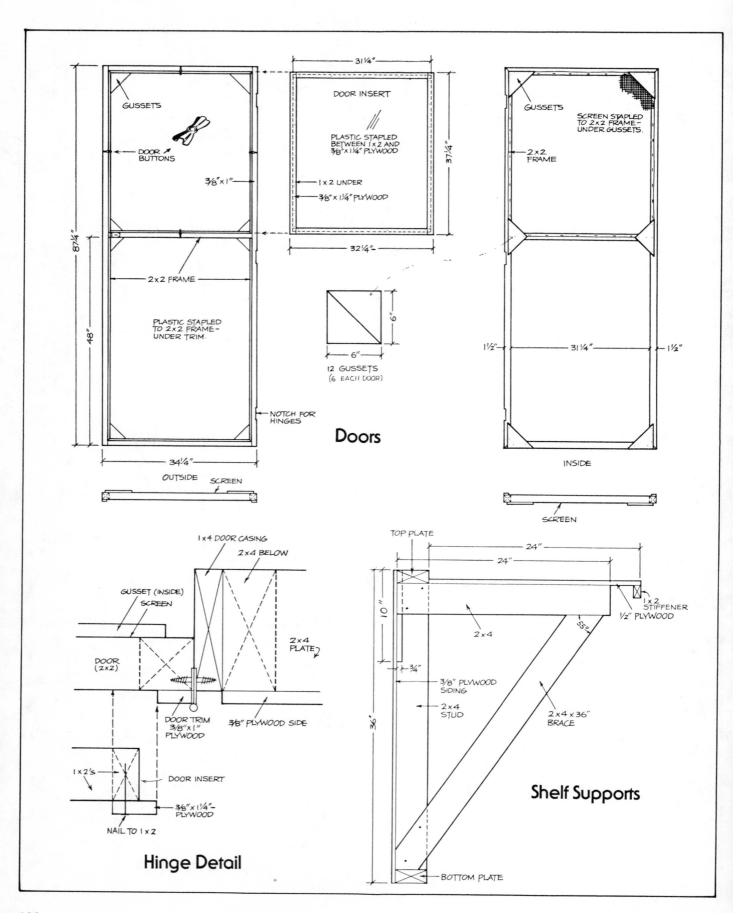

Doors

Hinge Detail

Shelf Supports

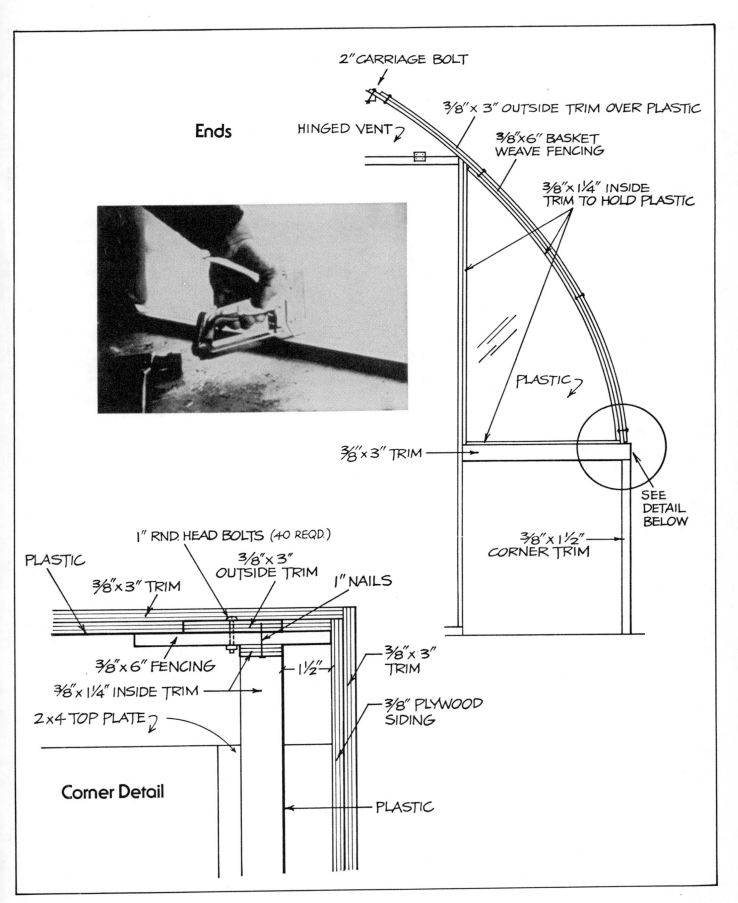

SECTION III: Specialized Needs

Panel Layout

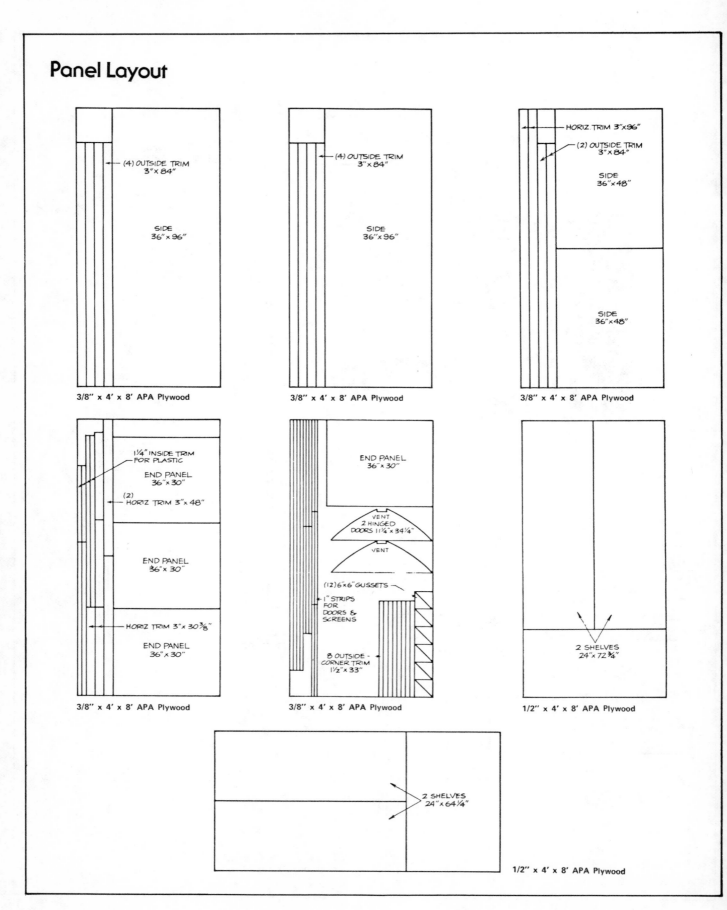

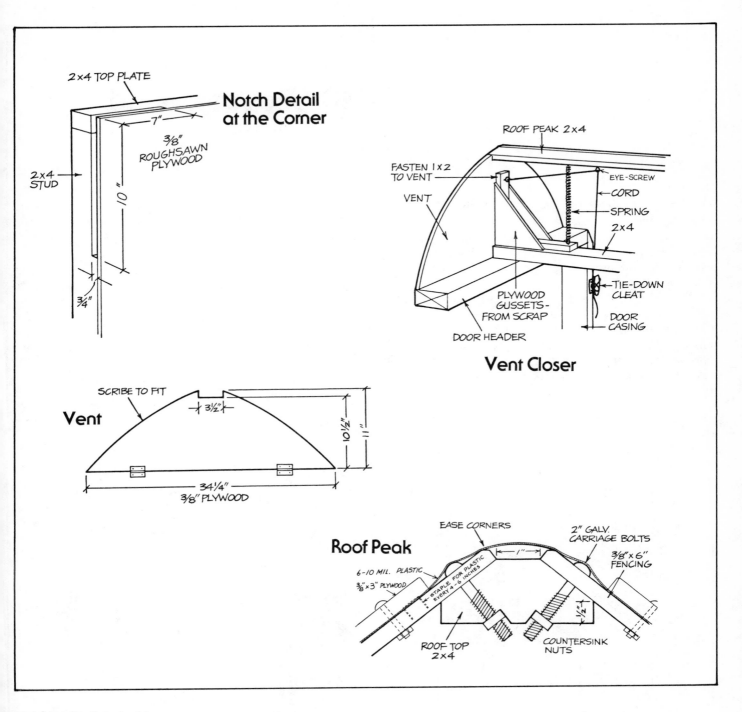

Finishing For Exterior Use

Since edges of plywood absorb moisture rapidly unless sealed, coat edges thoroughly with a high-quality oil-base exterior paint primer (if unit is to be painted) or a good water-repellent preservative (if unit is to be stained.)

For painting, always use a prime coat. Skimping on a primer can jeopardize the effectiveness of even the best top coats. Prime the unit just as soon as you can after assembly is complete. Use a primer that is compatible with the top coat used. Water-base acrylic latex paints, with companion nonstaining primers, are easy to use; they clean up easily and give excellent performance.

For rough or textured plywood, oil-base stains are recommended. Semi-transparent stain provides maximum grain show-through as well as maximum display of surface texture, but leaves little surface film. Opaque stain hides the grain pattern, but not the texture. Stains should be applied in one or two coats; as with paints, they give best performance if applied by brush. Brushing works the finish into the wood surface.

Whatever finishing method you use, paint or stain, always use top-quality materials, and follow the manufacturer's instructions.

13. Movable Shed

This movable shed was designed primarily for the grading of small fruits and vegetables, but it may be used for many other purposes: a poultry house, workshop, shelter, or storage area for small garden tools. Screening, shelving, or equipment would be useful, depending on the purposes for which the shed is to be used.

Construction is simple. Three pressure-preservative treated timbers, 4 x 4 x 13 ft., are used as skids. Wooden cross braces, 1 x 6 x 16 ft. 3 in., are notched into the skids and floor joists, and then secured with tenpenny common nails. The floor joists, placed 2 ft. on center, are fastened to the skids with steel framing anchors, metal straps, or ⅜-in. diameter bolts.

Corner posts are made of 2 x 4 double studs. These are nailed together with tenpenny common nails, staggered, and spaced 16 in. on center. These posts are fastened to the skids with steel framing anchors and metal straps.

The walls are constructed of 2 x 4 studs with 1 x 4 braces notched into them. This provides a strong, smooth surface to receive the exterior covering.

Roof joists are spaced 2 ft. apart and are secured with tenpenny and sixteenpenny common nails.

Exterior-type plywood is indicated on the drawings for the floor, walls, and roof, but other materials may be used.

Multi-Purpose Shed

Movable Shed

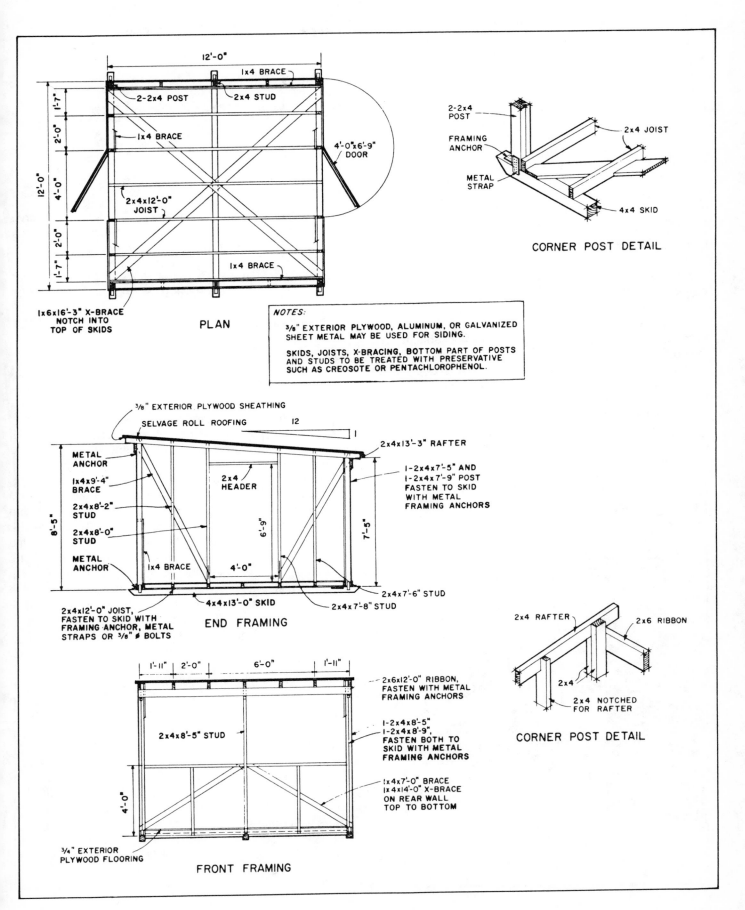

14. Housing for Horses

Barn

This three-stall barn for horses has been designed for use in all climates. An overhanging roof offers adequate shade in summer. In mild climates, the barn provides year-round protection against the elements. Insulation can be added for heat protection or for warmth in cold climates.

Before building, select a well-drained site, one that is sheltered from winter winds but open to summer breezes. An evergreen windbreak can provide an excellent shield. Face the outdoor pens toward the south or southeast.

Each stall has its own outdoor pen, which can be extended to any length desired. A 10-ft. gate can be installed between each pen partition as shown in the illustration. The gate can swing to either side and latches in position to close off the rear of the stall. Horse can be locked in the stall for bridling or holding while the pen area is being cleaned.

Build a Basic Horse Barn

Each 10 x 12 ft. stall has access to the inside area. Large sliding doors provide entry to the alley and the adjacent 10 x 8 ft. feed storage room. The 16 x 12 ft. grooming area permits easy access to the 8 x 8 ft. tack room.

A conventional door provides entry to the tack room. Inside, a unique pivoted door in the tack room makes the saddle racks easily available and saves steps when saddling or unsaddling. (See pp. 144-147.)

Portable Stable

This portable, one-horse stable was designed with emphasis on the economical use of materials.

For ventilation there are openings between the rafters at front and rear of the building, two hinged panels that open in the rear wall, and a dutch door. The windows in the sidewalls are optional.

The plan for the stable does not show a floor covering. However, well-tamped clay makes a very good floor for a horse stable and is preferable to a wood floor. The clay should be built up enough so that the floor will be several inches above the outside ground level. The walls can then be tilted up and secured to the pressure-treated timber base. The exterior-type plywood siding stiffens the walls, eliminating the need for let-in braces. The roofing is corrugated metal and translucent plastic. In areas subject to high winds, the stable should be securely anchored in place.

No interior wall lining is shown on the working drawings but may be desirable, depending on the temperament of the horse and the preference of the horseman.

CAUTION: Do not use paint containing lead on any area of buildings, fences or equipment that is accessible to livestock.

Indoor Exercise Area

Farmers who wish to establish recreational facilities on their property such as horse stabling and riding for added income should first check building codes, zoning laws, and other environmental regulations before starting construction. Sometimes there are definite limitations on the ownership, development, and use of property for these purposes, especially in urban and in developing areas.

The horse barn should be well planned, durable, and at-

Housing for Horses

Cutaway View of Barn

FEED STORAGE AREA   TACK ROOM   SWIVEL DOOR AND SADDLE RACK   GROOMING AREA   STALLS

SECTION III: Specialized Needs

Portable One Horse Stable

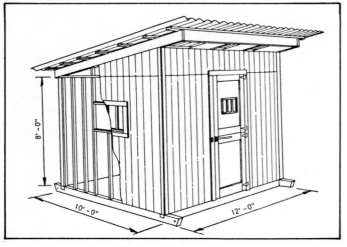

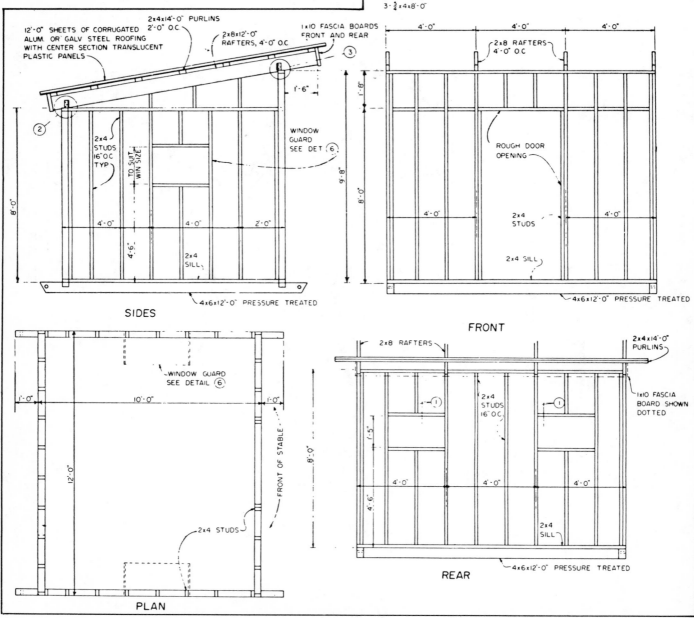

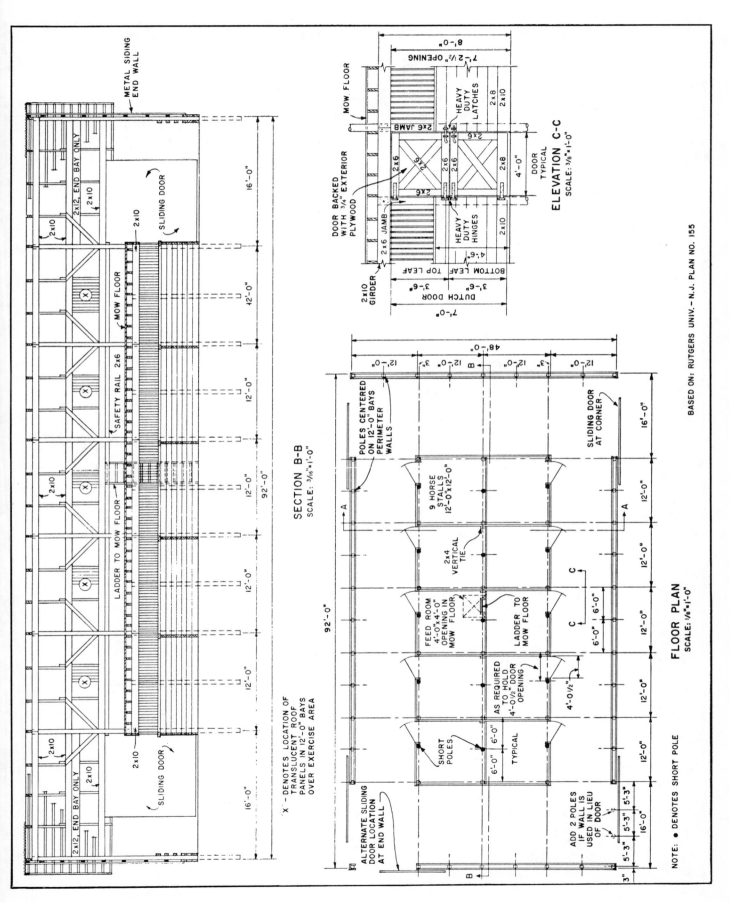

SECTION III: Specialized Needs

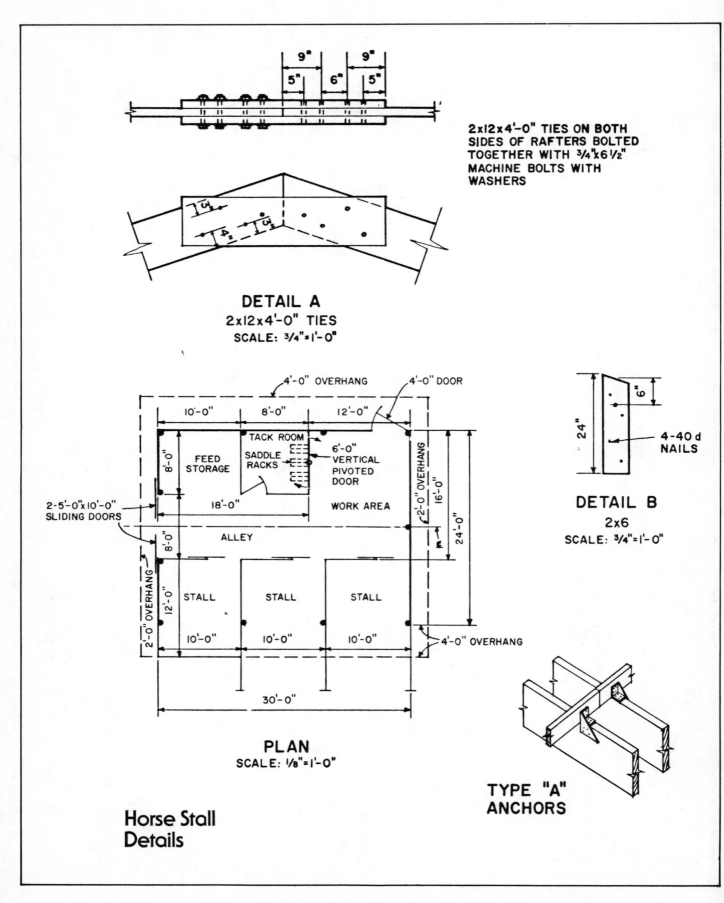

Horse Stall Details

Horse Barn Plans

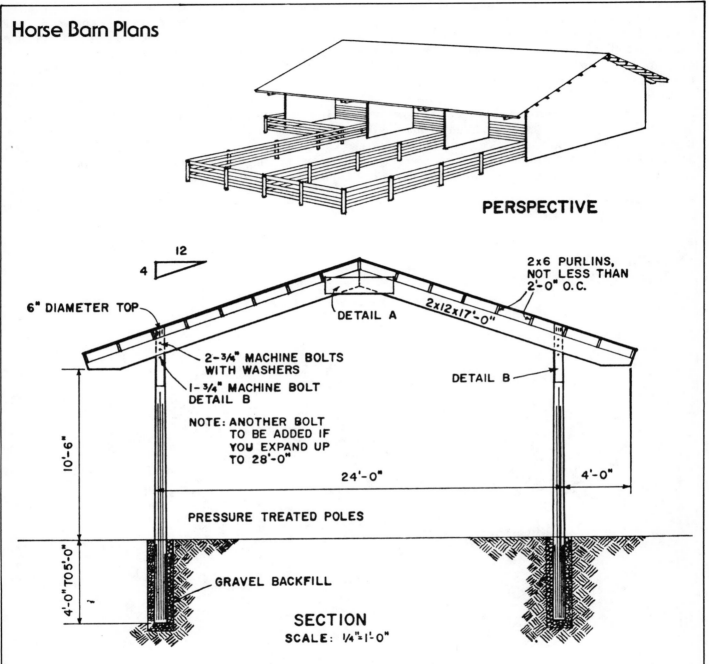

PERSPECTIVE

SECTION
SCALE: 1/4"=1'-0"

BILL OF MATERIALS FOR BUILDING WITHOUT WALLS OR PARTITIONS

- 12—2x12x17'-0" RAFTERS
- 3—2x12x16'-0" TIES DETAIL A.
- 36—2x6x12'-0" PURLINS & DETAIL B.
- 17—2x6x10'-0" PURLINS
- 8—6" DIAMETER x 18'=" OR 6x6x18'-0" PRESSURE TREATED POLES.
- 3—6" DIAMETER x 20'-0" OR 6x6x20'—0" PRESSURE TREATED POLES.
- 48—40d COMMON NAILS
- 102—PIECES OF TYPE A METAL ANCHORS FOR HOLDING PURLINS IN PLACE, OR 2x4x16" PIECES MAY BE SUBSTITUTED. NAILS FOR METAL OR WOOD ANCHORS.
- 1156 SQUARE FEET OF ROOFING.
- 68'—0" OF GUTTERS.
- 48 —3/4"x6½" MACHINE BOLTS WITH WASHERS
- 24 —3/4"x10" MACHINE BOLTS WITH WASHERS

END POLES REQUIRE A RAFTER FASTENED ON INSIDE OF POLE, ALL OTHERS REQUIRE TWO RAFTERS ON EACH SIDE OF POLE.
ADDITIONAL POST OR POLES AND LUMBER REQUIRED AS NECESSARY FOR PARTITIONS AND PENS.
PARTITIONS CAN BE MADE WITH 2" LUMBER, OR 1" EXTERIOR TYPE PLYWOOD.
TACK ROOM SHOULD BE MADE RODENT PROOF.
CONCRETE FLOOR IN STORAGE AREA IS PREFERRED.
USE HORSE PROOF AND RODENT PROOF GRAIN STORAGE.
A SLATTED FENCE PROVIDES A BETTER WIND BREAK THAN A SOLID FENCE.
THIS PLAN MAY BE LENGTHENED OR SHORTENED BY 10'-0".

tractive. Site selection is of major importance. Provide room for expansion and for specific details, such as water supply, manure disposal, electricity, windbreaks, feed storage, and loading and unloading access. The horse barn should fit your overall plan and harmonize with existing buildings. Consider nuisances that may affect your horses and your enjoyment of the facility.

This nine-stall horse barn is ideal for cooler areas or where an indoor exercise area is desirable. Illustrated below is the indoor arrangement, showing stalls, feed room, exercise area, and a mow floor over the stall area.

Inexpensive pole construction is used in the overall design, and poles are arranged to form the individual stalls. All poles, splashboards, and other wood in contact with the ground or with manure should be pressure-treated with a nontoxic preservative to a retention of 8 lbs. per cu. ft.

The 12 x 12 ft. stalls are constructed of 2-in. tongue and groove lumber with either ½-in. steel bars or with a commercial stall guard at the top. Tamped-clay floors are used for the stalls and exercise areas, and concrete floors for the combination feed and tack room.

Ventilation to prevent condensation is provided by continuous openings at the eave and ridge. Side panels and sliding doors can be opened for summer cooling.

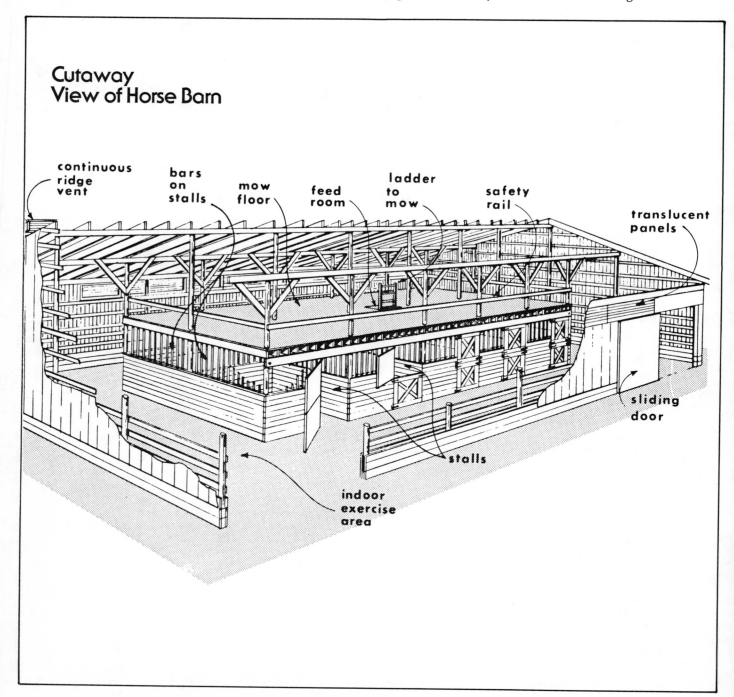

Housing for Horses

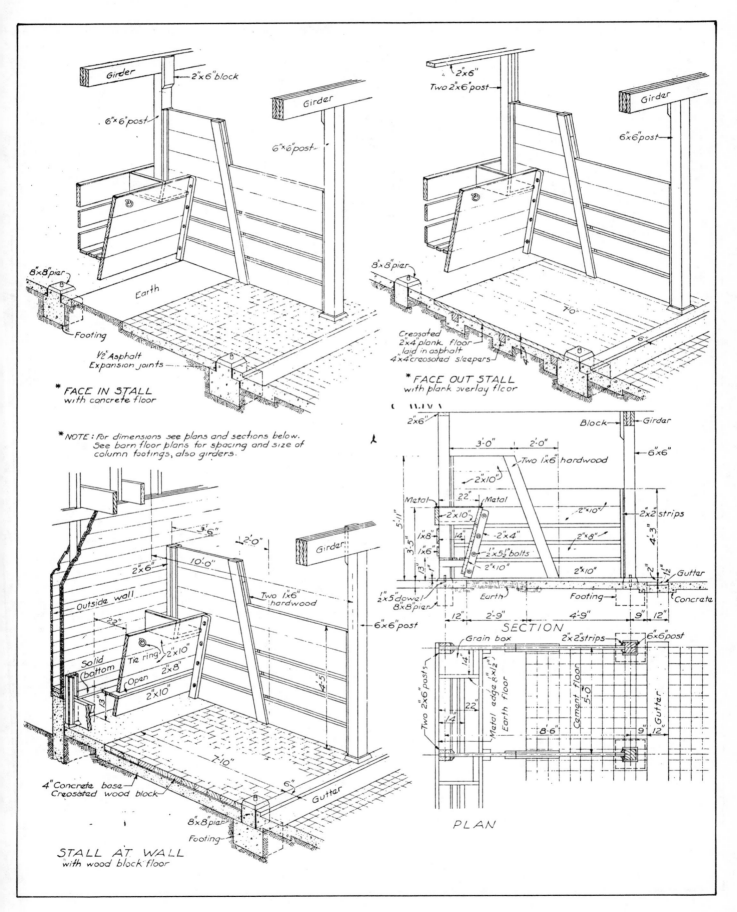

SECTION III: Specialized Needs

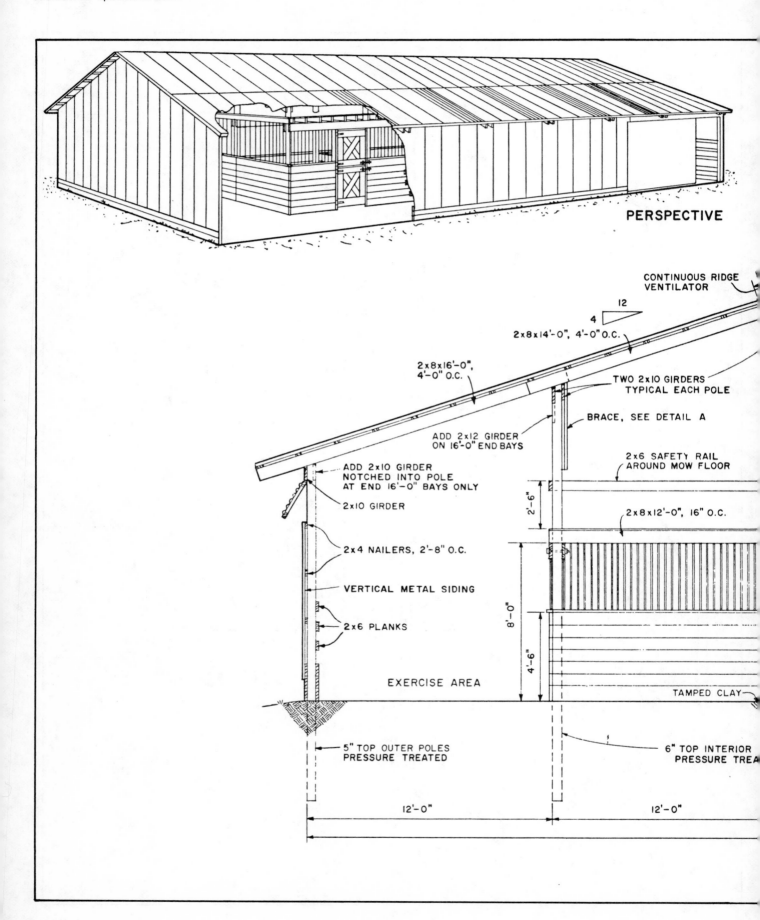

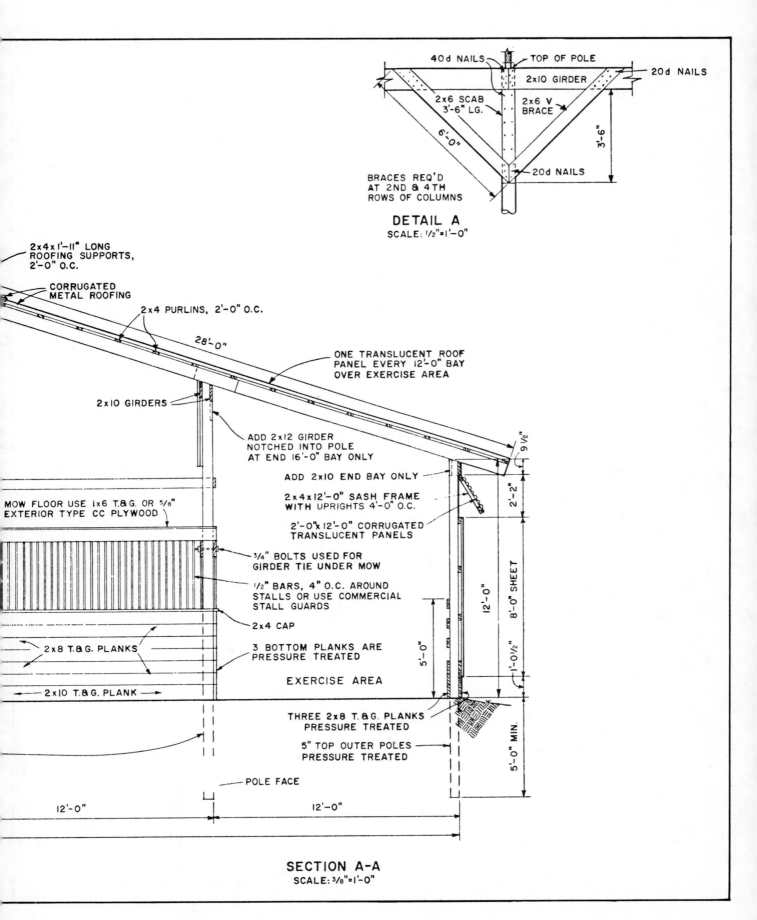

SECTION III: Specialized Needs

Construction Tip: Ventilation

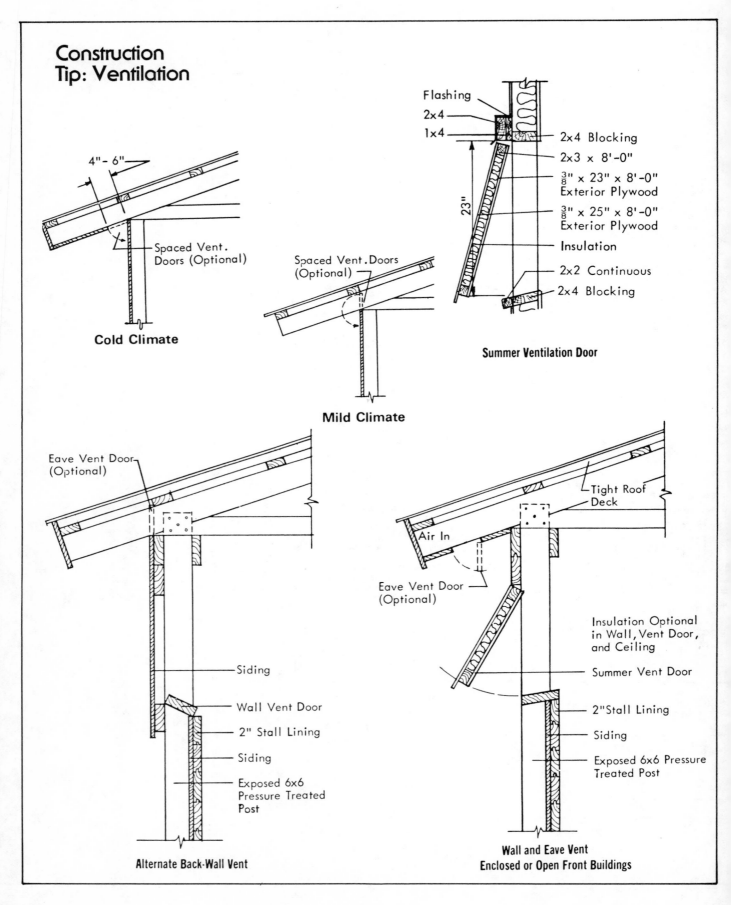

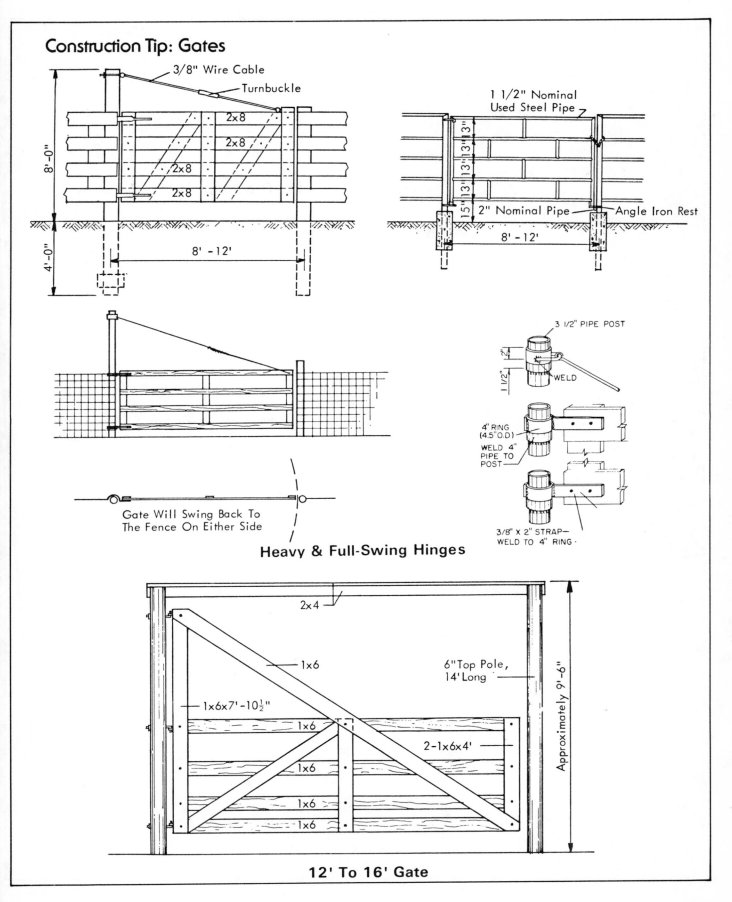

SECTION III: Specialized Needs

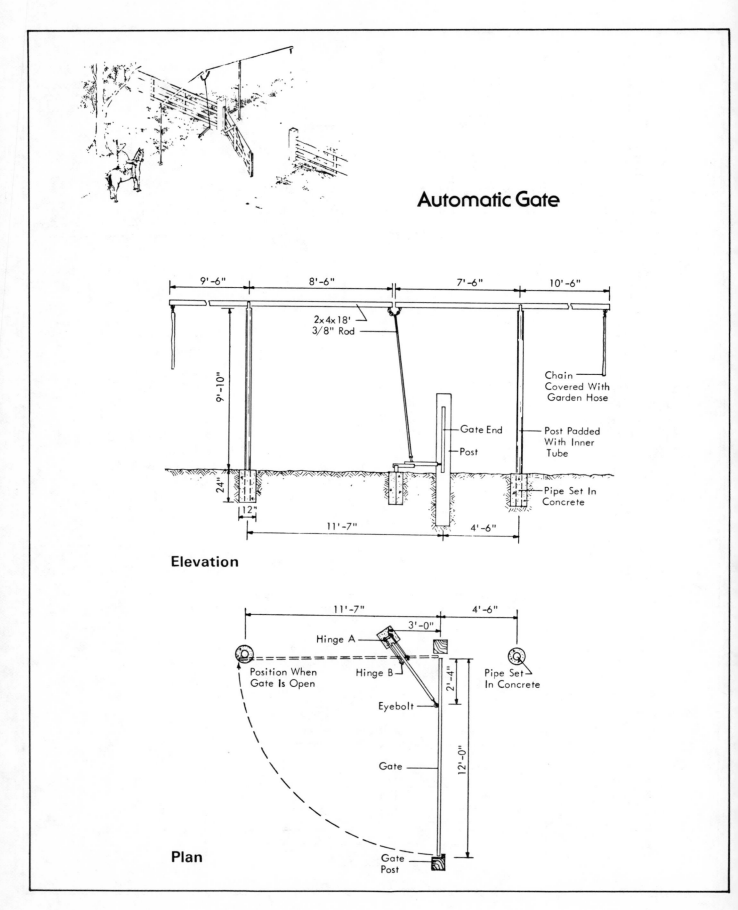

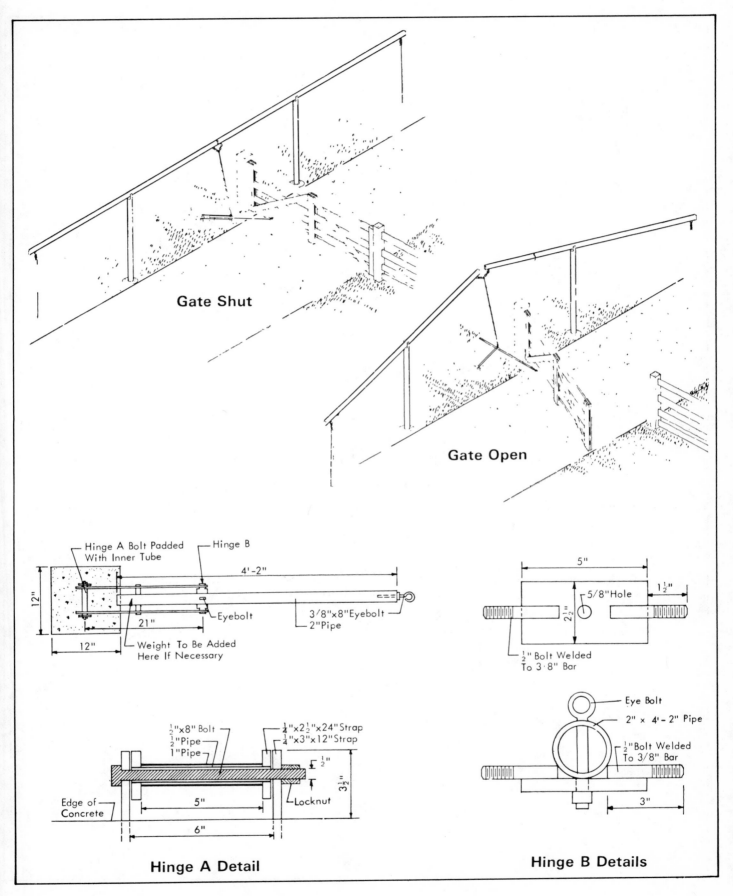

SECTION III: Specialized Needs

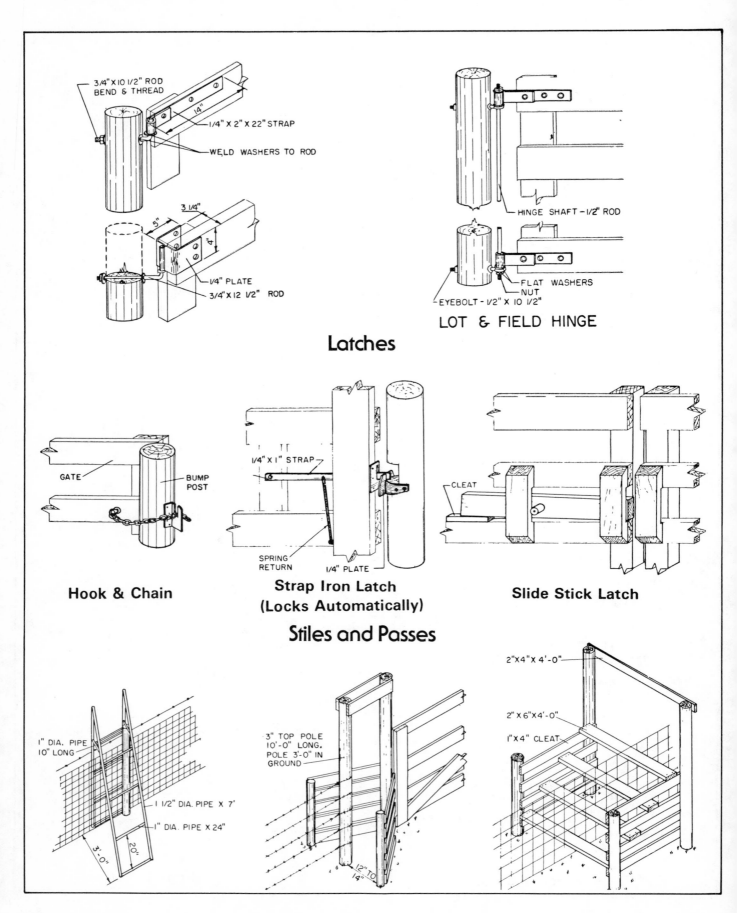

Latches

Hook & Chain

Strap Iron Latch
(Locks Automatically)

Slide Stick Latch

LOT & FIELD HINGE

Stiles and Passes

16. Cattle Shelters and Feeders

Adjustable Cattle Chute

This cattle chute was designed to meet the needs of those cattle ranchers and feedlot operators who handle various sizes of animals.

The movable side of the chute is hinged at the bottom to angle irons set 15 in. into concrete. The bottom width of the chute is 24 in. The top of the adjustable side of the chute can be moved in and out. Depending upon the sizes of animals being handled, the top can be set in the desired position by inserting a pin in the proper hole in the steel adjusting bar, which slides through the stationary wall of the chute.

The floor of the chute, made of concrete 3 in. thick, slopes 2 in. to one side to provide easy removal of the manure and liquid.

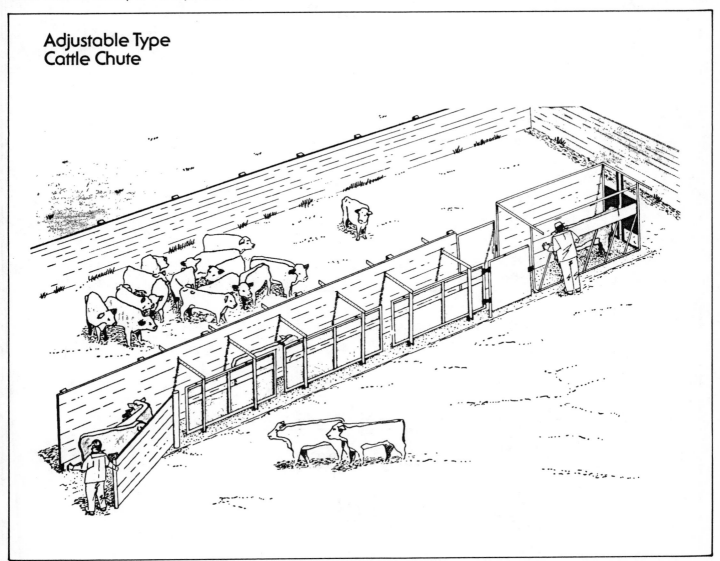

Adjustable Type Cattle Chute

SECTION III: Specialized Needs

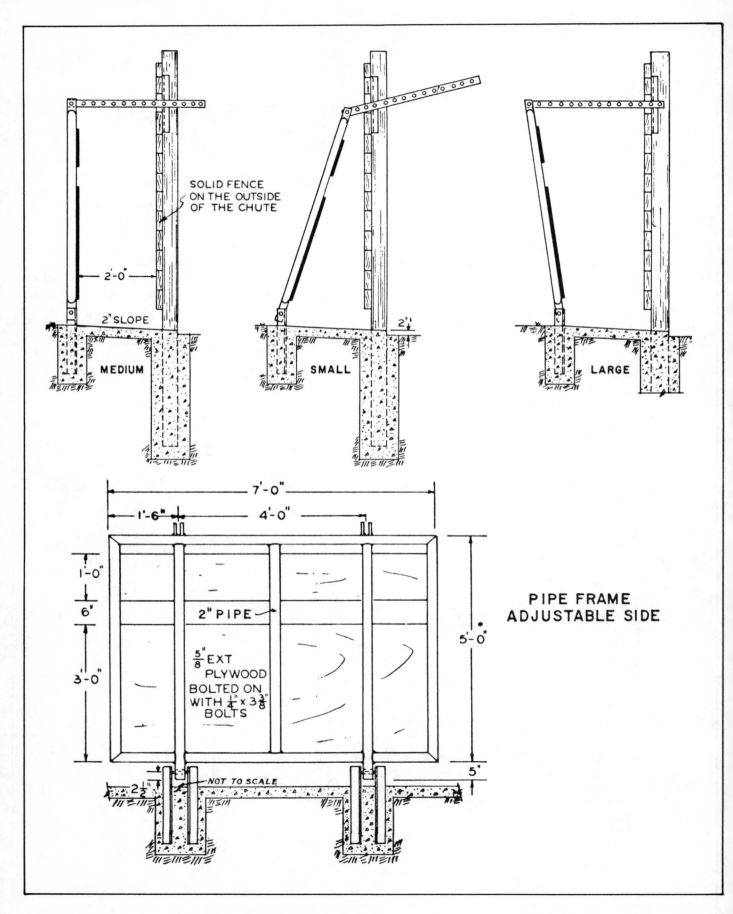

Wagon Rack for Silage-Feeding

This wagon rack may be built without flooring and adapted to a standard wagon bed.

In selecting a chassis to carry the bed, care should be taken to keep the height of the feeding bunk within 30 in. of the ground.

Consider the size of the cattle to be fed when ordering the wheel-and-axle assembly, as it may be desirable to use a drop-axle for lowering the feeder. Short breeds, such as Jersey or Aberdeen Angus, feed better if the height is not more than 27 in.

The running gear selected must also be adequate for handling heavy loads, as the feeder box with center diverter has a maximum capacity of 117 cu. ft.

Use of high-flotation rims and tires are suggested to improve mobility and to reduce the possibility of bogging down in soft ground.

When full, the 16-ft. wagon holds about 1.7 tons of silage—enough to feed 88 head of cattle for one day.

By using all four sides of the wagon, the feeding space will accommodate approximately 60 head of cattle.

If the wood parts are treated with a good nontoxic preservative, like pentachlorophenol or copper naphthanate, the wagon should resist weathering.

CAUTION: If you paint the feeder, do not use paint containing lead.

SECTION III: Specialized Needs

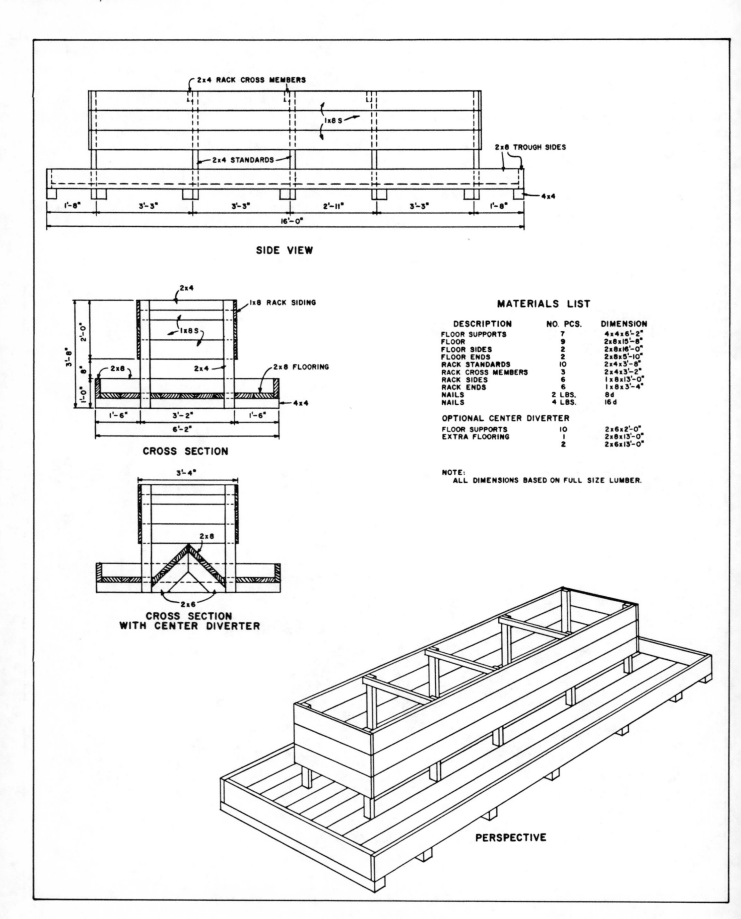

16. Shelters and Feeders for Swine

Hog Feeder

Effects on the carcass value of swine forced to eat while standing on their hind legs were studied in a recent environmental project at the University of California, Davis, Calif., in cooperation with the U.S. Department of Agriculture. Pigs were fed from elevated troughs to investigate the possibility that the departure from their normal eating position would exercise more of the higher-priced muscles of the ham and possibly the loin.

Two troughs, similar to the 8-ft. trough illustrated, were constructed with the upper edge of the front lip of the feeding compartment about 38 in. from the ground. Trough height could be changed by raising or lowering the pigs' feeding platform or raising the feeder up or down. The feeding compart-

SECTION III: Specialized Needs

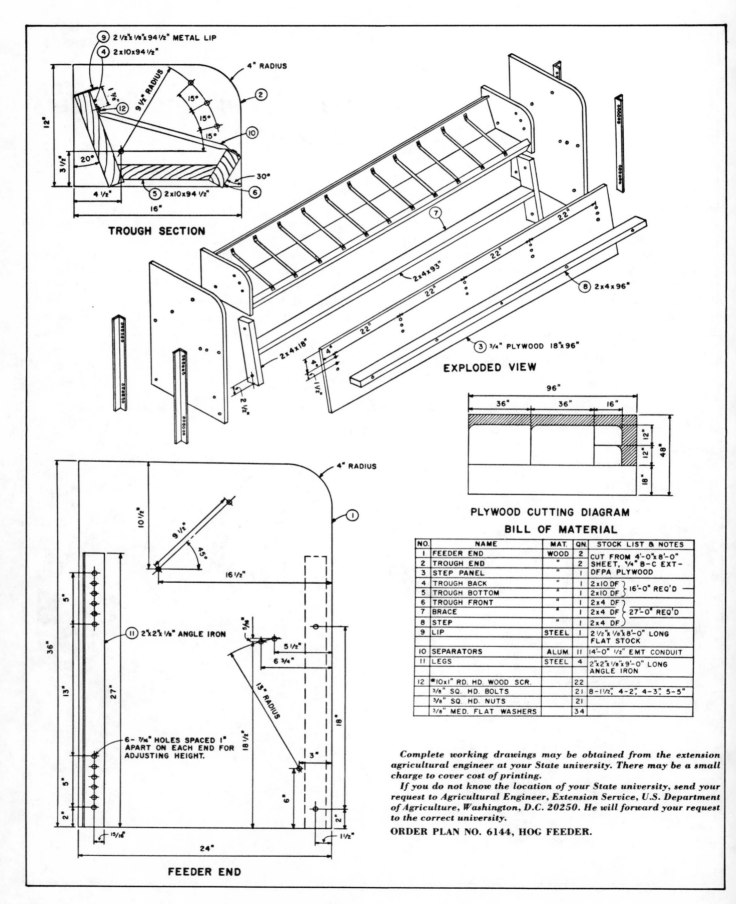

Complete working drawings may be obtained from the extension agricultural engineer at your State university. There may be a small charge to cover cost of printing.

If you do not know the location of your State university, send your request to Agricultural Engineer, Extension Service, U.S. Department of Agriculture, Washington, D.C. 20250. He will forward your request to the correct university.

ORDER PLAN NO. 6144, HOG FEEDER.

Swine

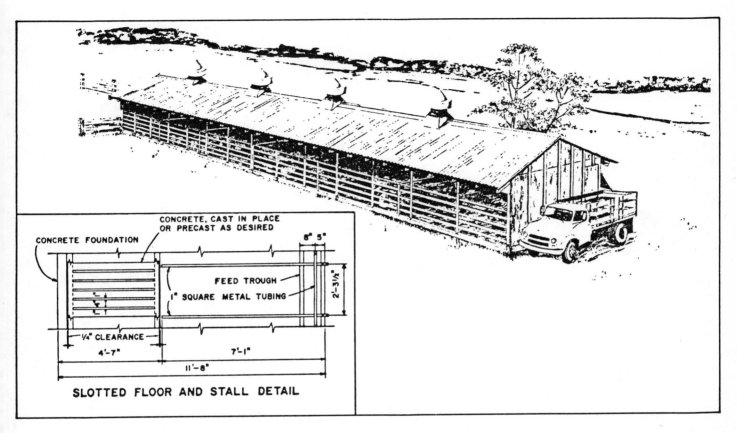

SLOTTED FLOOR AND STALL DETAIL

ment was so constructed that feed tended to stay along the front edge. The compartment could be rotated to change the height of the feeder, as well as the slope of the bottom of the compartment.

The test results showed that forcing pigs to stand on their hind legs to feed may decrease their weight, but may increase the percentage of ham, or ham and loin.

House for Dry Sows

This building has an open front, partially slotted floor. It is a well-arranged structure for dry sows.

About one-third of the floor is used for the slotted area. The solid part slopes ¾ in. per ft. into the manure tank under the slotted floor.

The feed can be handled in several ways. The most common way is to use a feed cart to deposit the feed into the trough for each sow.

There should always be some movement of natural ventilation air regardless of the outside weather. This building depends entirely on this type of ventilation. Open to the south or east, for minimum winter exposure, this building provides a series of 3 x 8-ft. doors along the back (north or west) wall for adequate air movement during the hot weather. Ventilators, as shown, also help insure adequate air movement. Some producers close a portion of the open front with plastic curtains during cold and windy weather. A screened opening is provided at the eaves plate (front and back wall) for natural ventilation.

Farrowing House

This 24 x 48 ft. farrowing house made of 8 x 8 ft. preframed insulated wall panels was designed by the American Plywood Association. These panels may be purchased prebuilt, built on-site, or shop-built before erection. This type of construction is popular in the Midwest, with lumber dealers supplying either preframed panels or the complete building package from their yard.

Once the floor and manure pits have been poured, the preframed panels can be erected on the 2 x 4 sill and secured by adding the 2 x 4 top plate.

Full-confinement swine housing is the most modern method of swine production. With a structure of this type, the farmer endeavors to regulate all the environmental needs of the swine. This unit is suited to the climatic conditions of the Midwest, but with proper selection of insulation and a ventilation system, it can be adapted to most areas of the country.

Two alternate ventilation systems which provide summer cooling and winter air movement to keep the moisture levels low inside the structure are shown on the working drawings.

Two slotted-floor arrangements made of either concrete or wood slats are shown on the working drawings.

The plywood farrowing-stall detail is compatible with this house. Stall sides are cut from a single 4 x 8 ft. sheet of ¾-in. exterior-type plywood. The ends of the stall and divider partitions are also made of plywood.

SECTION III: Specialized Needs

Housing for Dry Sows

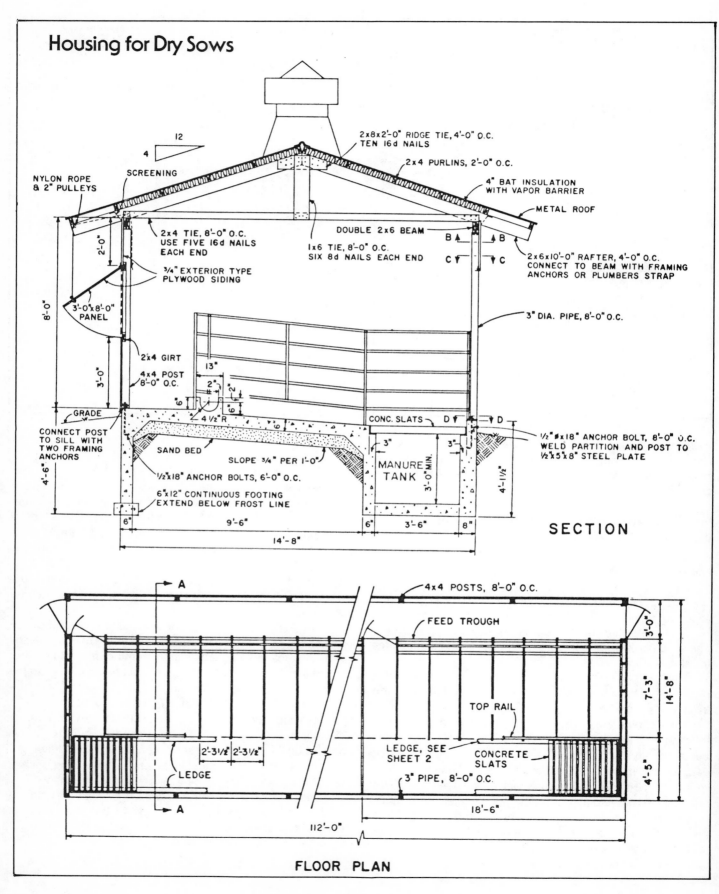

Farrowing House

Design Details

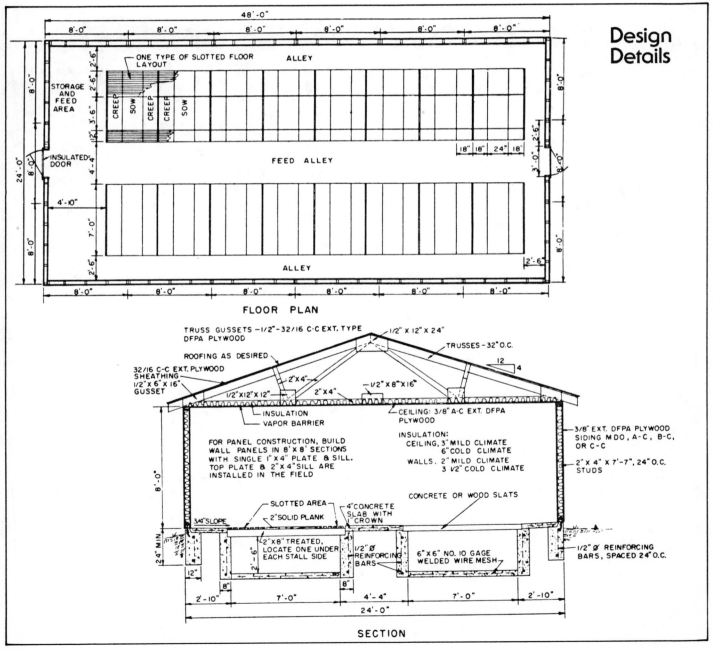

17. Poultry Accessories

If you want to keep a few hens for egg production, consider the plan followed by many producers: buying 20-week-old started pullets for the laying house. This is a good way of getting started with some birds.

Another possibility is to contact a commercial egg producer who is selling his flock of old hens. Some of these hens will do a good job of laying eggs. You should plan to mold these birds (allow them to regenerate their feathers, and rest for 8 to 10 weeks). This should enable them to come back into production and give you large, good-quality eggs.

A small chicken house wih ventilation inlet slots between the windows.

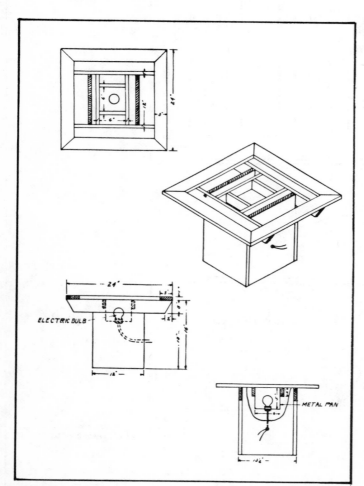

A simple plan for a water stand with light bulb to prevent water from freezing.

A plan for a small feeder.

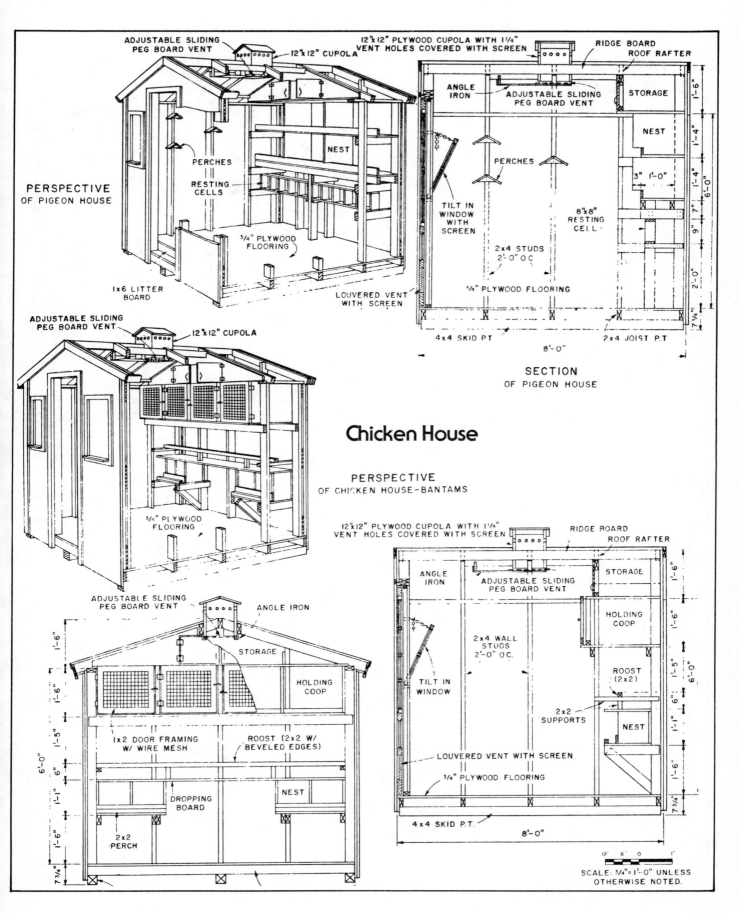

SECTION III: Specialized Needs

Brooder and Laying House

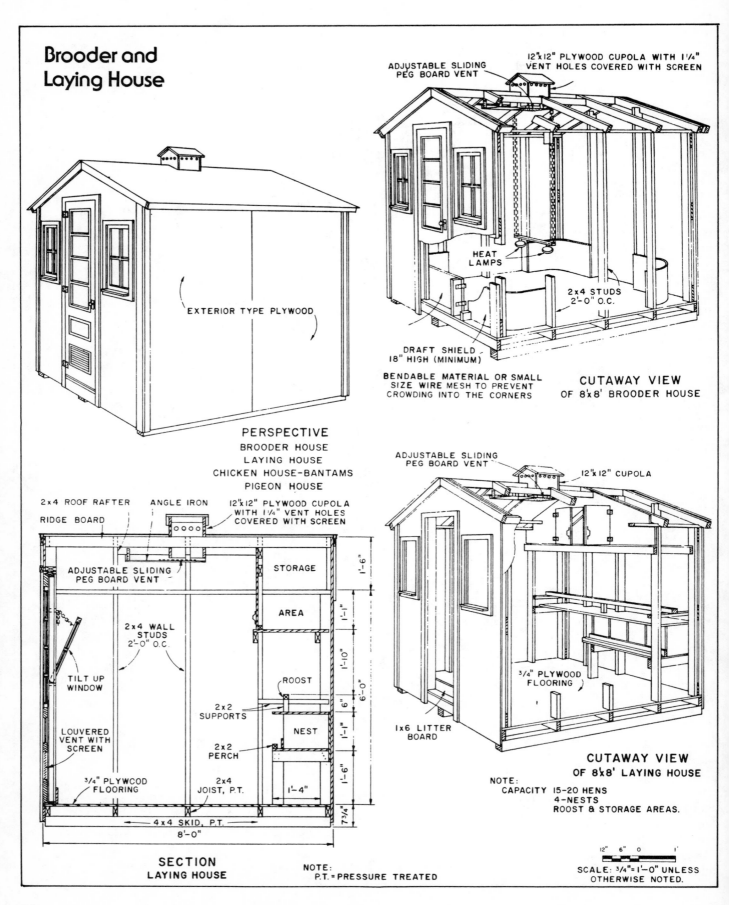

SECTION 4: Maintenance

18. Fences

Fences may be used to protect or divide property, to improve its appearance, or to confine animals. Whatever their purpose, they should be planned carefully. This is especially important on farms where they may represent a large investment and their location and arrangement may affect production efficiency.

Permanent fences, those intended to last for many years, should be well constructed and made of good materials. Temporary fences, those intended to stay in place only a short time, need not be so sturdily constructed and may be made of less expensive materials.

Selection of Fencing

The kinds of fences commonly used on farms include board, woven wire, barbed wire, combination woven wire and barbed wire, cable and electric. Board, rail, and chain link fences are the most popular kinds for rural-home lots.

The basic considerations in selecting fencing are its purpose, the cost, and your preference.

Productive farmland is often fenced with woven wire or a combination of woven wire and barbed wire. Marginal, cutover, or other less productive or nonproductive land, is usually fenced with less expensive materials such as one or more strands or barbed or smooth wire.

Woven wire fences, or a combination of woven wire and barbed wire, are commonly used to confine livestock. Different styles or designs are recommended for the different kinds of animals. Special styles or designs of wire fencing are also available for confining poultry.

Board fences are also used to confine livestock and are very popular on horse farms. When painted white or whitewashed, strong board fences can be very attractive as well as functional.

Cable fencing is excellent for feedlots and similar areas where livestock are closely confined. This open-type fencing allows unrestricted air circulation through the lot, resulting in maximum cooling of the animals in warm weather.

Electric fencing is convenient for temporary usage where more permanent fencing would be too expensive or perhaps undesirable. Woven wire and board fences may also be temporary.

White-painted board fences or redwood fences make an attractive framing for a farmhouse or rural home. Or for a more rustic appearance, a two- or three-rail fence might be used.

Chain link fence is commonly used to protect front or back yards or to confine pets in yards.

Preparation

Before erecting a fence, you may have to lay out the fence line, or clear it, or both.

Laying Out a Fence Line

To lay out a fence line on level ground, set a stake at each end of the proposed fence line and station another person at one of the ends. Starting from that end, set a stake every 100 ft., with the second person verifying the alignment of the stakes with the two end stakes.

To lay out a fence line over hills where you cannot see the other end stake, set two stakes on top of the hill where both can be seen from both end stakes. Line up the two stakes, first with one end stake, then with the other. You may have to move one or both of the stakes several times to obtain satisfactory alignment.

Clearing a Fence Line

Fence lines should be cleared of trees, brush, stumps, rocks, old fencing, and other obstructions that might interfere with construction of the fence or detract from its appearance. The easiest and quickest way to clear a fence line is to use a bulldozer or a bulldozer blade mounted on a tractor. With this equipment, you can knock down small trees and old fencing, clear brush, and level high spots and fill in low spots all in one operation.

Large trees can be cut down or pulled down with a tractor. If you pull them down, use a heavy rope, cable or log chain—long enough for you to be safe from the falling tree. Be careful of the dangerous recoil if the rope, cable, or chain should break.

Once trees, larger brush, and old fencing have been removed, you can plow down or turn under small brush and grass with a disc harrow or a field cultivator. (Shown are some of the hand tools available for clearing small growth. Always wear leather gloves to protect your hands.) If you clear a strip through woods or thicket, make it wide enough to allow you

SECTION IV: Maintenance

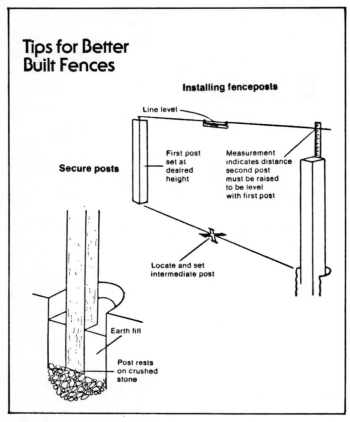

Drawings courtesy of Koppers Co.

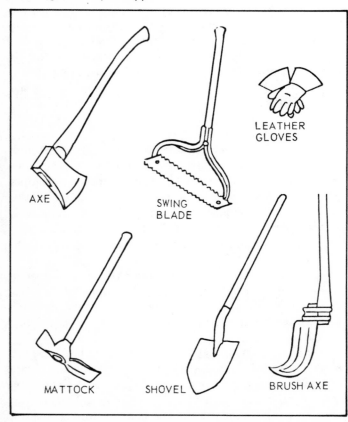

Handtools for clearing brush from land.

to distribute the posts and unroll the fence. A wide strip can later serve as a roadway or fire lane.

When replacing old fencing, you may want to use some old, but sound, posts. The easiest way to remove them from the ground without damage is to use the hydraulic lift on a tractor or an A-frame.

Old wire is usually not worth saving; never leave it in fence corners or other places where it may become a hazard to livestock. One way to dispose of it is to stake it in ditches to help prevent soil erosion.

Fenceposts

Fenceposts may be made of wood, steel, or concrete. Concrete posts are used mainly in farm fencing. Considerations in determining the kind, size, and number of posts to use include: (1) the availability and cost of the different kinds; (2) the kind of fence you plan to erect, how strong it needs to be, and how long you want it to last.

Wood

Wood posts can be bought in most areas and are comparatively low in cost. In many areas, the farm woodland may be a good source of wood posts.

For permanent fencing, you should use the most durable kind of wood posts available, or better still, use pressure-preservative treated posts.

The durability of untreated wood posts, even of the more decay-resistant kinds, depends largely on the heart wood content. Whether bought or cut, untreated wood posts should be of mostly heartwood. Untreated sapwood of any wood species will usually rot in 1 to 3 years.

The probable life expectancy of different kinds of untreated wood posts of mostly heartwood is:

Kind	Years
Osage Orange	25 to 30
Red Cedar and Black Locust	15 to 25
Sassafras	10 to 15
White Oak, Blackjack Oak, and Cypress	5 to 10
Southern Pine, Sweetgum, Hickory, Red Oak, Sycamore, Yellow Poplar, Cottonwood, and Willow	2 to 7

Osage Orange, Red Cedar and Black Locust posts may no longer be available in some areas. If less-durable posts are used, they should be treated with a good wood preservative to protect them against decay and insect damage. Depending on the kind of preservative used and the method of application, treating posts can extend their life 10 to 30 years.

Pressure-treated posts—posts treated with preservative by commercial process—are usually more durable than farm- or home-treated posts. Creosote or some other preservative is forced into the wood under pressure at a rate of 6 to 8 lbs. per cu. ft.; such posts are available in most areas.

Farm or home methods of treating wood posts with preservative include: hot and cold bath, cold soak, end diffusion, and double diffusion. These methods—and the precautions to be followed—are described in Farmers' Bulletin 2049, reproduced in the appendix (the material is out of print at the Office of Agriculture).

Brushing on wood preservative is not recommended for wood posts. The wood will not absorb enough of the chemical to give effective protection against decay.

Wood posts can usually be bought in lengths of 5½ to 8 ft. and in diameters of 2½ to 6 in. or larger. Posts 5 in. or larger in diameter are generally used for anchor posts (gate, corner, end, and braced-line posts). Line posts for straight, open-field woven wire fences are sometimes as small as 2½ in. in diameter, but a minimum diameter of 3½ in. is recommended. Four- or 5-in. posts should be used for barn lots, corrals, and other confined areas and in sandy and wet soils.

Height of the fence and the depth of post setting determine the length of posts required. Anchor posts are usually set 3 to 3½ ft. in the ground and line posts are usually set 2 to 2½ ft.

Steel

Steel posts offer a number of advantages. They are lightweight, fireproof, extremely durable, and easily driven into most soils. Also, they will ground the fence against lightning when in contact with wet or moist soil.

The more common kinds of steel posts are line posts and corner posts. Steel posts are usually sold in lengths of 5, 5½, 6, 6½, 7, 7½, and 8 ft.

Stock crowding against the fence tend to force steel posts out of line. Anchor plates, bolted, clamped, or riveted to the posts as shown, help keep the posts firmly in the ground.

If you live near oilfields, boiler factories, or repair shops, you may be able to buy used pipe at reasonable cost for use as fenceposts. The pipe should be at least 1¾ in. in diameter for line posts and larger for anchor posts. For heavy anchor posts, use pipe 6 to 8 in. in diameter, filled with concrete.

Concrete

Concrete posts are used mostly as anchor posts (gate, corner and end posts) in farm fencing. However, where a fence angles slightly, use of a concrete post will prevent the wire from pulling line posts out of line.

If well made, concrete posts can be extremely durable, lasting 30 years or longer. If not well made, they may start to crumble in a few years.

Concrete line posts are usually 4 in. square and are cast several at a time in forms of the desired length. Anchor posts are usually larger and are cast in place.

Woven Wire Fences

Woven wire fencing is made in different classes. Those for farm and general use include field or stock fencing, poultry-garden fencing, chick fencing, and wire netting. Chain link fence, a more stylized form, is designed mainly for home lots.

In most classes of woven wire fencing, you have a choice of fencing weights, protective coatings on the wire, and style or design.

Weight

The weight of woven wire fencing is determined by the gage, or size, of the line, or horizontal wires. The lower the gage number the larger the wire, and the larger the wire the stronger and more durable the fencing.

Field or stock fencing, for example, comes in four weights:

Gage of line wires:	Top and bottom	Intermediate
Lightweight	11	14½
Medium weight	10	12½
Heavy weight	9	11
Extra heavy weight	9	9

The stay (vertical) wires in the fencing are usually of the same gage as the filler (intermediate line) wires and may be spaced 6 or 12 in. apart.

Protective Coatings

Most woven-wire fencing is either zinc coated (galvanized) or aluminum coated. However, chain link fence also comes with a vinyl resin coating, which makes a more attractive fence.

The coating on zinc-coated fencing may be Class 1, 2, or 3. The class number indicates that the fencing has at least the following amount of galvanizing per square foot of wire surface:

	Oz. of zinc coating per sq. ft. of wire surface		
Wire gage	Class 1	Class 2	Class 3
9	0.40	0.60	0.80
10	0.30	0.50	0.80
11	0.30	0.50	0.80
12½	0.30	0.50	0.80
14½	0.20	0.40	0.60

The thicker the zinc coating the more corrosion resistant the fencing. The class number will be indicated on the tag on the roll of fencing.

The coating on aluminum-coated fencing is usually about 0.25 oz. per sq. ft. of wire surface. This is not usually indicated on the tag, however.

Under the same climatic conditions, aluminum-coated fencing could be expected to resist corrosion three to five times longer than zinc-coated fencing with the same thickness of coating. However, in rural areas, if the coating were broken so that the wire was exposed to the air, then the zinc would give better protection. While both metals act as a "sacrificial agent"—they corrode instead of the wire—in rural atmospheres an oxide film tends to form on the aluminum

SECTION IV: Maintenance

coating, limiting its ability to protect the wire. Any wire fencing will resist corrosion longer in a dry climate than in a humid area or in an industrial atmosphere.

Styles or Designs

The styles or designs of woven wire fencing are designated by a three- or four-digit number; for example, 932 or 1155. The first one or two digits indicate the number of line wires in the fencing and the last two indicate fence height in inches; style 1155, for example, has 11 line wires and is 55 in. high.

Shown are the five most commonly used styles or designs of field or stock fencing: 1155, 1047, 939, 832, and 726, all combined with barbed wire.

Cattle and horses. Use Fence A or B. The single barbed wire at the top prevents the animals from mashing down the fence.

Hogs. Use Fence C, D, or E without the barbed wires above the woven wire. The barbed wire below the woven wire discourages the animals from crawling or rooting under the fence. Styles 939 and 832 are available with a barbed bottom wire.

Style 726. This is Fence E without barbed wire; is convenient for temporarily confining hogs while they hog down corn.

Sheep. Use Style 832 or 726. These are Fence D or E without barbed wire; barbs may tear the fleece on sheep.

To protect sheep from dogs and coyotes, extend barbed wire along the top to discourage dogs from jumping the fence. Add an apron of woven wire 18 in. wide along the ground to prevent predatory animals from burrowing beneath the fence.

For cattle, horses, hogs, and sheep all in the same field, use fence A, B, D, or E.

Poultry-garden fencing. Available in two standard styles—2158 and 1948; chick fencing comes in three styles—2672, 2360, and 2048.

Wire netting of 1- or 2-in. mesh is made in nine heights ranging from 12 to 72 in. The netting is commonly used for fencing small poultry yards, cages, poultry house windows, and tree guards. The 1-in. mesh wire is recommended for confining baby chicks, turkey poults, and goslings.

Construction

Follow these safety precautions, as fence construction always involves the risk of injury:

(1) wear heavy leather gloves, boots or high shoes, and tough, close-fitting clothing;
(2) never use a tractor to stretch woven wire or barbed wire fencing, because while up on the tractor you may not be able to tell when the fencing has been stretched to the breaking point (if the wire should break, you could be injured seriously by the recoil of the clamp bar, chain or fencing);

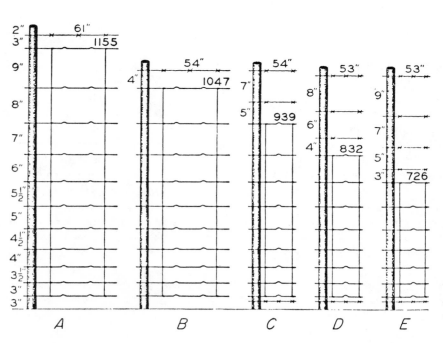

Standard style or designs of woven wire fencing combined with barbed wire. Stay (vertical) wires are spaced 12 in. in fences A and B and 6 in. for D, D, and E.

(3) carry staples, nails, or other fasteners in a metal container or in an apron—not on your person—and under no circumstances carry them in your mouth;
(4) when stretching woven wire or barbed wire, stand on the opposite side of the post from the wire and stretcher unit;
(5) if you handle preservative-treated posts, do not rub your hands or gloves on your face or other parts of your body; some people are allergic to the chemical.

Setting Posts

Setting untreated wood posts in concrete is not recommended. The post may shrink from the concrete, leaving bility and anchorage provided by the concrete may justify its use. Setting the treated posts in concrete is also recommended when the fence may be subject to heavy pressure or to wind strain.

Dirt-set wood posts may be set in pre-dug holes or driven into the ground. Power post hole diggers or post drivers can save much time and labor in setting posts. Post drivers are available that can drive posts up to 8 in. in diameter.

For a stronger fence, set steel anchor posts (gate, corner, end, and braced-line posts) in concrete. Dirt-set types of steel posts are available, but they will not provide quite so strong an assembly. Steel line posts should be driven directly into the ground, not set in concrete.

Corner and End-Post Assemblies

Corner- and end-post assemblies are the foundation of a fence. If one fails, the whole fence or a section may come down.

Shown are the different types of wood assemblies. The double-span assemblies have more than twice the strength of the single-span assemblies and only half the horizontal and vertical movement under heavy loads.

Single-span assemblies may be used for fence lengths up to 10 rods. For fence lengths of 10 to 40 rods, use double-span construction. Over 40 rods, use double-span construction plus braced-line posts.

Minimum sizes recommended for the components of the assemblies are:

Single spans
Corner post . . . . .6-in. diameter
Brace post . . . . .5-in. diameter
Brace . . . . .4-in. diameter
Tie .2 double strands of No. 9 gage wire

Double spans
Corner post . . . . .5-in. diameter
Each brace post . . . . .4-in. diameter
Each brace . . . . .4-in. diameter
Each tie . . . . .2 double strands of No. 9 gage wire

Height of the fence and the depth of post setting determine the length of posts needed. The posts should be set at least 3½ ft. in the ground.

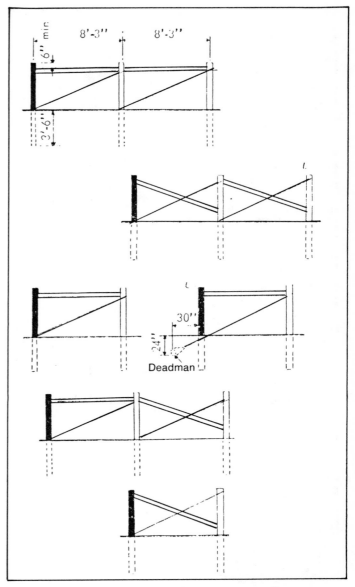

Wood corner- or end-post assemblies. Corner or end post is shaded.

Here are the steps in constructing single-span assemblies. Repeat as necessary for double-span assemblies.
(1) Dig the holes for the anchor and brace posts, spacing them 8 ft. apart.
(2) Set the anchor post, but not the brace post. Tamp the soil firmly as you replace it around the post. Lean the top of the post 1 in. away from the direction of fence pull so that it will straighten to a plumb position when the fence is stretched.
(3) Stand the brace post in its hole and fasten the wood brace to both posts. Use dowel pin construction for a strong assembly.
(4) Set the brace post, tamping the soil firmly as you replace it around the post.
(5) Attach the brace wire as shown and splice the ends together. Tighten the wire by twisting it with a strong

SECTION IV: Maintenance

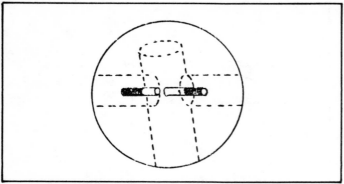

Use dowel pins to connect wood brace to wood corner or end post or brace post.

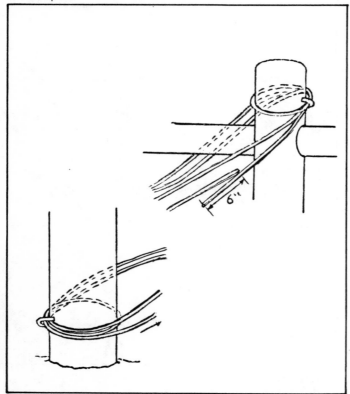

Fastening wire brace or tie-in wood corner- or end-post assembly.

stick or rod. Leave the stick or rod in place so that you can adjust the tension when necessary.

For construction of a steel corner- or end-post assembly, both the post and the braces should be set in concrete.

(1) Dig the hole for the post, at least 3½ ft. deep. For a corner post, make the hole 20 in. square at the bottom and 18 in. square at the top. For an end post, make the hole 20 in. square at the bottom but 18 x 20 in. at the top with the long dimension parallel to the fence line. With these hole dimensions, the concrete piers for the posts will have a pyramidal shape and be more stable.

(2) Attach the metal braces to the post.

(3) Holding the post plumb in its hole, mark the holes for the braces on the ground. The braces should enter the concrete pier 6 in. below the ground surface and should extend 6 in. into the concrete. Therefore, the center of the holes will be closer to the post than where the braces touch the ground.

(4) Dig the holes for the braces, making them at least 18 in. deep and 20 in. square at the top and bottom. At least 8 in. of the concrete pier should be below the frost line.

(5) Holding the post plumb in its hole, place the concrete around it and the braces. Tamp the concrete as you pour it, and slope it slightly away from the post and braces to drain water.

Line Posts

Line posts are usually spaced 14, 16, or 20 ft. apart for field fencing and 10 to 16 ft. apart for home lot fencing. Closer spacing may be necessary if the ground is uneven or if you need a stronger fence. Near the corner or end of the fence, you may have to shorten the spacing between the posts to equalize it.

Set wood line posts 2 to 2½ ft. in the ground and steel line posts 1½ to 2 ft. Stretch a cord, rope, or wire between the two anchor posts to serve as a guide in aligning the posts.

When set in low places, line posts should be weighed down, set in concrete, or provided with subsoil cleats to help hold them in the ground.

Do not set line posts in a gully or stream where they could be washed out by a heavy flow of water. If the fence line crosses a narrow gully or stream, stretch the fence straight across from a well-secured post on each bank. Barbed wire can be stretched below the fence to prevent stock from crawling under it.

If the gully or stream is wide, terminate the fence section on the one bank with an end-post assembly, and start a new section on the other bank. Install a floodgate across the gully or stream to restrain livestock.

Braced Line Posts

In fences 40 rods or longer, braced line posts should be used every 20 yards. Construction is the same as for the corner or end post assemblies shown as A and E, except that a second brace wire is used to take the fence pull in the opposite direction.

Stretching and Attaching the Fencing

Woven wire fencing should be stretched and attached in sections running from one anchor post to the next. Anchor posts include gate, corner, end, and braced-line posts.

Do not attach the fencing to concrete-set posts until the concrete has thoroughly hardened.

In a combination woven wire and barbed wire fence, attach the woven wire first. Instructions for stretching and attaching barbed wire are given in the section on barbed wire fences.

For the best appearance, fasten the fencing to the "outside" of the posts. But if the fence will be subject to

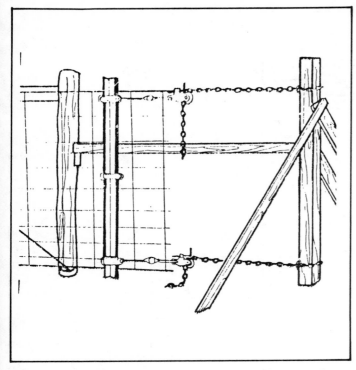

Fence stretcher in position; note braced "dummy" post at right.

pressure—from livestock, for example—fasten the fencing to the "inside" of the posts. Here are the basics.

1. Starting about 2 ft. ahead of the anchor post, unroll the fencing to the second line post, and stand the fence roll on end.
2. Remove one or two stay wires from the end of the fencing to free enough length of each line wire to wrap around the post and splice on itself.
3. With the next stay wire against the post, staple the fence to the post at the desired height.
4. Starting with the middle wire, wrap each line wire around the post and back on itself. Make five wraps around the wire, using a splicing tool.
5. Unroll the fencing to the next anchor post.
6. Set a dummy post for attaching the stretcher unit 4 to 8 ft. beyond the second anchor post. Brace the post as shown.
7. Prop the fencing against the line posts with stakes. Support it at the top, about a foot from every third or fourth post, and on the opposite side from the stretcher unit.
8. Attach the stretcher unit to the fencing and to the dummy post. Use a single-jack unit for fencing up to 32 in. in height and a double-jack unit for higher fencing. Attach the jack of a single-jack unit to the center of the bar and the jacks of a double-jack unit so that the wires are divided equally between the jacks.
9. Stretch the fence slowly so that the tension will be evenly distributed over the entire length. Check the fencing during the stretching operation to make sure that it is riding free at all points. Continue the stretching until the tension curves in the wire are straightened out about one third. Do not overstretch the fence. If the tension curves are straightened out too much, the fence will lose some of its springiness.
10. Staple the line wires to the anchor post.
11. Loosen as many stay wires as necessary to complete step 12, and slide them toward the stretcher.
12. Starting with the middle wire, cut each line wire and wrap it around the post and back on itself four turns. Do every other wire, working toward the top and bottom, until all are done. Leave the top wire until last.
13. Starting at the end farthest from the stretcher, fasten the fencing to the line posts. Fasten the top wire first, then the bottom wire, then every other wire until all are fastened.

Contour Fencing

Contour fencing may be required on terraced land or in strip cropping. It calls for slightly different construction.

Post spacing. Post spacing must be reduced wherever there is much curve in the fence line to keep the posts from overturning. To determine the proper spacing, stake out a smooth curve along the terrace or contour strip, spacing the stakes about 14 ft. apart. At any one point of curvature, select three consecutive stakes and stretch a string between the first and third stakes. Measure the distance from the center stake to the string and space the posts as below.

Distance from center stake to string, in.	Post spacing, ft.
Inches	Feet
4 or less	14
5 to 6	12
7 to 8	10
9 to 14	8
15 to 20	7

Repeat the above procedure wherever the curvature of the fence line appears to change noticeably. Then check by eye to see that no post is out of line of a smooth curve. With a smooth curve, the fencing will pull equally against each post.

Setting posts. Lean the top of the post about 2 in. toward the side to which the fencing will be attached. When the fence is stretched, the post will straighten to a plumb position.

Installation. On curves, attach the fencing to the outside of the posts so that it will pull against the posts. This may mean that the fencing will be first on one side of the posts then on the other.

Contour fencing can usually be stretched in 20 per 4-rod sections. However, on sharp curves, it may be necessary to stretch it in 10-rod sections. Where the curvature of the fence line changes materially, it is a good idea to start a new section at the sharpest point on the curve.

Stretch a section of contour fencing from the end having the least curvature, so that the end having the most curvature will have the least tension.

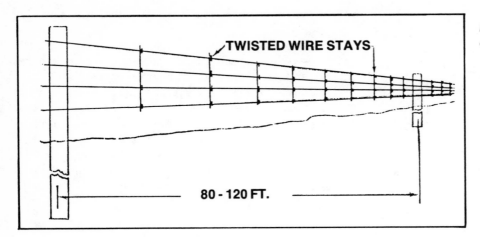

Barbed wire suspension fence, commonly used a cross or boundary fencing on cattle ranges.

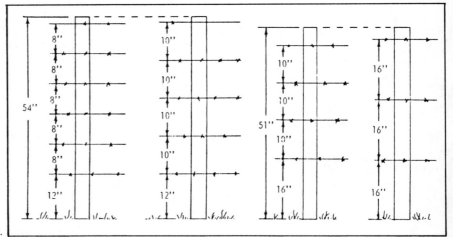

Common spacings of wires in barbed wire fences.

Barbed Wire Fences

Barbed wire is used both in conjunction with other fencing and as fencing itself. Its use in combination with woven wire fences and for electric fences is covered in the sections on those kinds.

Design

Shown is the usual wire spacing in three- to six-strand barbed wire fences. As few as two strands are sometimes used to fence large cattle ranges in the Western States.

Barbed wire suspension fences are often used as cross fencing and boundary fencing on large cattle ranges. They consist of four to six strands of the wire supported by posts spaced 80 to 120 ft. apart. Twisted wire stays, spaced about 16 ft. apart, hold the wires apart.

When cattle come in contact with a suspension fence, it sways back and forth, beating against the cattle and discouraging them from trying to go through it.

Of the standard barbed wire commonly available, the 12½-gage wire with 2-point barbs is the most widely used for cattle ranges. For smaller fields, where cattle may subject the fence to considerable pressure, four-point barbs may be more effective. The lighter 14-gage wire is commonly used for temporary fencing.

You can also buy high-tensile barbed wire, which is stronger and more durable than the comparable sizes of standard wire. The 13½-gage high-tensile wire, for example, has a breaking strength equal to that of the 12½-gage standard wire.

Barbed wire, like woven wire, comes with a protective coating of either zinc or aluminum. Thickness of the coating is the same as on comparable sizes of woven wire. Under the same climatic conditions, aluminum coated wire would be more durable than zinc-coated wire.

Installation

Barbed is especially dangerous to work with because of the barbs. Follow the safety precautions given already and take any others necessary to avoid injury.

Strong anchor post assemblies are essential if the barbed wire does not have tension curves. Any pressure on the fence will be transferred directly to the posts.

Installation is the same whether the wire is used in combination with other fencing or as separate fencing. Unroll, stretch, and fasten one line at a time. In a combination fence, attach the barbed wire below the woven wire first. Then attach the wires above the woven wire, starting with the lowest one and working upward.

The basic steps are given here.

1. Fasten one end of the wire roll to the anchor post, leaving enough wire free to wrap it around the post and splice it later (step 2). If the anchor post is a gatepost, remove the barbs from the wire to be wrapped around the post to prevent injury to persons or animals using the gate.
2. Wrap the wire around the post and splice it onto itself, 3½ to 4 turns.
3. Unroll the wire along the ground to the next anchor post. Unroll it straight off the roll, not off the side.
4. Set up a dummy post about 8 ft. beyond the second anchor post and brace it. If you are erecting a combination woven wire and barbed wire fence, you can use the dummy post set up to stretch the woven wire fencing. Attach a fence stretcher or a block and tackle unit to the dummy post, and attach the wire to the stretcher unit.
5. Stretch the wire until it is fairly tight. Be careful not to stretch it so tightly that it breaks.
6. Fasten the wire to the anchor post.
7. Remove the barbs from a sufficient length of wire to wrap around the anchor post and splice on itself.
8. Cut and untwist one of the two strands of the wire; the other strand will maintain the tension. Wrap the cut strand around the post and back on the wire 3½ to 4 turns. Leave enough space between each turn to interwrap the second strand. Cut the second strand, wrap it around the post, and splice it on the wire.
9. Fasten the wire to the line posts.

Cable Fences

Cable fencing consists of heavy galvanized cables attached to metal posts or running through holes drilled through metal or wood posts. When well constructed of good materials, it makes strong, durable fencing.

The fencing is excellent for feedlots and similar areas where cattle are closely confined because it allows unrestricted air circulation through the area, resulting in maximum cooling of the animals in warm weather. Combined with woven wire fencing or barbed wire, the fencing can be used to confine hogs and sheep as well as cattle.

Shown is the construction where the cables run through holes drilled through wood posts. Each cable is attached to the anchor post by a spring assembly. The cable is stretched with a block and tackle until the spring begins to open and is then clamped around the next anchor post. When necessary, adjust the tension of the cable by tightening or loosening the spring.

For details on additional constructions, obtain instructions from fence manufacturers or dealers.

Board Fences

Board fences are very popular both on farms and for home lots. Strong board fences are excellent for confining livestock and, when painted white, can be attractive as well as functional. Avoid the use of lead paints on fences.

Designs can be varied by increasing or decreasing the number of boards, their width, or the spacing between them.

Strong fences are required for corrals, feedlots, and similar areas where livestock are closely confined and may subject the fence to considerable pressure. For such areas, planks 2 in. or more in thickness, 8 or more in. in width, and 10 to 16 ft.

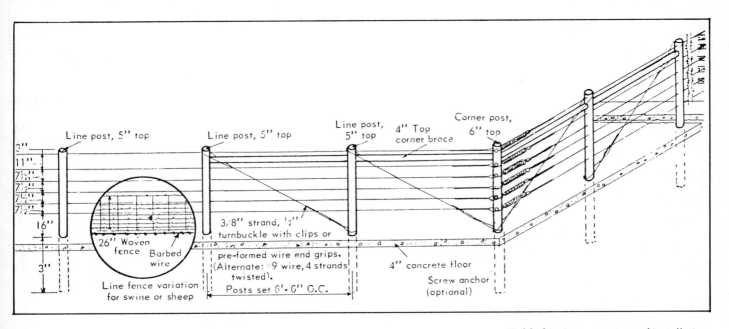

Cable fencing—one type of installation.

SECTION IV: Maintenance

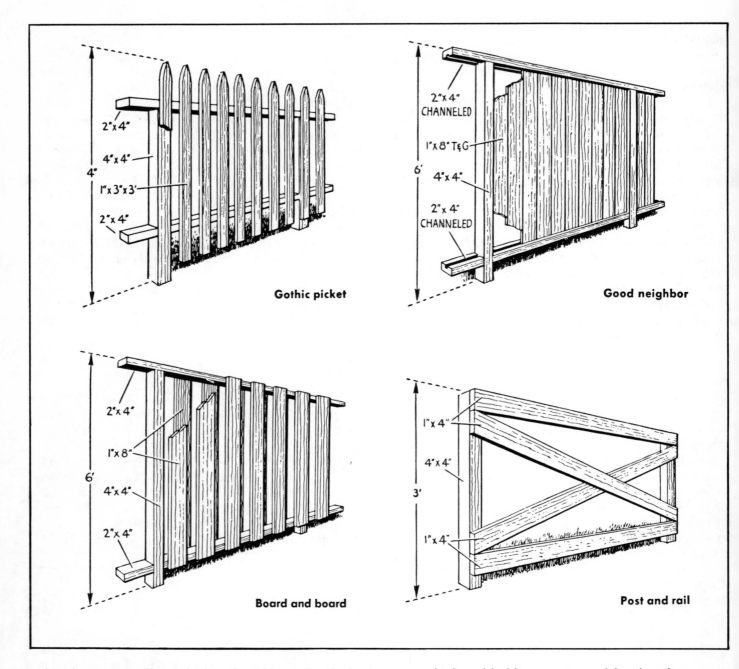

in length, are generally used. They should be well spiked or bolted to substantial posts, spaced 5 or 6 ft. apart.

For field or home lot fences, boards 1 in. thick (¾ in. dressed), 6 in. wide (5¾ in. dressed) and 6, 8, 10, 12, or 16 ft. long are generally used. When the posts are spaced the usual 8 ft., the use of 16-ft. lengths will save labor and make a stronger fence.

Top and side fascia boards are used mainly for appearance, but they afford some protection and strength to the fence.

Lumber Quality

Minimum requirements in lumber for fences and gates are moderate bending strength, medium decay and weather resistance, high nail-holding power, and freedom from warp. Woods combining these properties to a high degree include cypress, Douglas fir, western larch, southern yellow pine, redwood, and white oak.

Other woods that weather well but have a small tendency to warp, and less strength and nail-holding power are: cedar, northern white pine, ponderosa pine, western pine, sugar pine, chestnut, and yellow poplar.

Woods that are strong, hard, and high in nail-holding power but have a greater tendency to warp and do not weather so well as the preceding group include: beech, birch, red gum, maple, red oak, and tupelo.

Eastern hemlock, western hemlock, white fir, and spruce are intermediate in the properties between the two preceding groups.

The No. 1 or No. 2 softwood or No. 2 Common hardwood grades of lumber should be used to achieve stronger, more durable gates and fences.

Preservative Treatment

Fences and gates will last years longer if the wood is treated with a good preservative. If not treated, the wood may soon start to decay at joints or any place where moisture is held.

Commercially treated lumber may be available in most areas. Pressure-treated wood will last longer than that treated by other methods.

You can treat the wood yourself by soaking it in a preservative solution. If you will paint the fence, use a clear preservative that will not bleed through the paint, such as penta (pentachlorophenol) or copper naphthenate in light oil (mineral spirits or kerosene). If you will not paint the fence, you can use creosote, penta, or copper naphthenate in heavy (fuel) oil.

The lumber should be thoroughly seasoned before being treated with preservative. Green lumber will not absorb enough of the chemical for good protection against decay.

Cut the boards to the desired length before treating them. Soak them in the solution for at least 15 minutes, but preferably for 1 hour for each inch of thickness. The longer the boards soak, the more preservative they will absorb and the longer they will resist decay.

While not the best method, you can apply preservative with a brush. Small quantities of ready-to-use wood preservative are available from farm-supply stores. The preservative should be flooded-on, and the treatment should be repeated every few years for best results.

Construction

Construction of board fences is essentially the same regardless of design. Given here are the more important details in building a four-rail fence, using 4-in. posts, 16-ft. boards, and top and side fascia boards.

Slope (saw) the tops of the posts slightly toward the side to which the boards will be fastened so that the top fascia boards will slope toward that side.

At the start of the fence and at corners, set the first line post 7 ft., 10 in. from the gate or corner post, center to center. Space the other line posts 8 ft. apart, center to center. The first boards should extend across the face of the anchor post to the center of the first or second line posts. Subsequent boards are nailed center to center on the posts.

Near corners or the end of the fence, you may have to shorten the spacing between posts to make it come out even. For the best appearance, however, make the last panel as nearly full length as possible.

In the first or first two panels at the start of the fence and at corners, use 16-ft. boards for the top and third rails and 8-ft. boards for the second and fourth rails. With this arrangement, only two joints will fall on any one post and you will have a stronger fence.

For the best appearance, fasten the boards to the "outside" of the posts. However, if the fence will be subject to pressure—from livestock, for example—attach the boards to the "inside" of the posts.

Nail the boards to the posts with three ring-shank or screw-shank nails, staggered to avoid splitting the board. Or, for a stronger fence, hold the boards in place with cleats bolted to the posts. This method makes it easier to remove the boards quickly if necessary.

In the first panel at the start of the fence and at corners, use an 8-ft.-long top fascia board. Then, the top fascia-board joints and the top fence-board joints will fall on different posts. The top fascia should overlap both the top fence board and the vertical fascia board.

Caution: If you saw, trim, or bore preservative-treated boards, you may expose untreated or inadequately treated wood. Apply preservative if necessary to prevent decay.

Electric Fences

An electric fence consists of one or more electrically charged wires supported by, but insulated from, wood or metal posts. Either smooth wire or barbed wire may be used. A controller, commonly called a fence charger, is required to regulate the amount and timing of the current through the wire.

Electric fencing is commonly used to confine cattle and horses, but it can also be used to control hogs and sheep. A single charged wire along the top or side of a wood or other kind of fence will deter stock from crowding the fence and breaking it down. This may make it possible for you to use an old permanent fence in poor condition.

Electric fencing is low in cost, economical to operate, and easy to erect and move around. But to be completely effective the fence must be kept in continuous operation and the livestock must be trained to respect it.

The fencing should be installed and operated in accordance with the National Electrical Code, state and local regulations, and the manufacturer's directions. The fence charger and other equipment used should carry the label of Underwriters Laboratories (UL) or the Industrial Commission of Wisconsin. Fence chargers not carrying these labels may cause injury and death to humans and livestock.

Fence chargers are usually designed to operate on 110-120 volt power. Battery operated units are available for use where electric power is not available.

Approved fence chargers emit the current intermittently, not continuously. The "on" time is usually $1/10$ of a second 45 to 55 times a minute. The shock is sharp, but short and harmless.

Locate your fence charger in a building or otherwise protect it from the weather.

SECTION IV: Maintenance

Gateways

Gateways are an important part of both farm fencing and home-lot fencing. They should be located and designed to facilitate travel into and around the property.

Farm gateways include cattle guards and floodgates as well as conventional gates.

Working drawings of plans for construction of conventional gates, cattle guards, and floodgates are available from the extension agricultural engineer at your state university.

Gates

Gates usually receive as much or more wear than the fence. Permanent gates, therefore, should be strong, well hung, well braced, and made of good materials. For barnlots, paddocks, and similar areas where livestock are closely confined, gates must be strong enough to withstand frequent bumping by the animals.

Aluminum gates weather better than wood or steel gates, but they do not hold up as well when subjected to pressure from livestock or damage from farm machinery. You can greatly extend the life of wood gates by treating the wood with a good preservative.

Farm gates are usually 52 to 54 in. high. A 10- or 12-ft. width is usually adequate for livestock. For farm machinery, the gate should be 14 ft. or more in width.

Cattle Guards

Cattle guards permit the passage of vehicles through the fence line while restraining livestock.

A guard strong enough for heavy vehicles may be built of heavy planks set on edge, steel rails, small L-beams, or pipes 2 to 3 in. in diameter. It should be at least 8 ft. wide and, to prevent severe jolting of the vehicle as it is driven across, the members should be spaced not more than 3 in. apart.

If the guard is intended to restrain sheep or goats, there should be no smooth strip across the top of it. Sheep and goats will walk across a strip as narrow as 2 in.

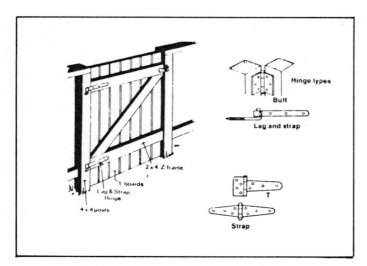

Shown here are the basics of gate building, including 4 types of hinges. If you wish to remove the gate at times, use the lag-and-strap style hinge, which permits the gate to be easily lifted off without unscrewing the strap or hinge.

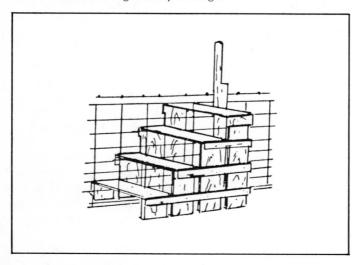

A stile should be used where a fence must frequently be climbed.

Cattle guards allow vehicles to pass through fence line, but restrain livestock.

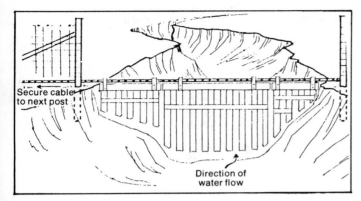

Floodgates are used where fence lines cross wide streams and gullies.

The pit beneath the cattle guard should be 12 to 18 in. deep. Use crankcase oil to control weeds and mosquitos in the pit.

Floodgates

A floodgate may be used to restrain livestock where the fence line crosses a wide stream or gully.

Shown is one kind of floodgate. Strong, well-secured end-post assemblies are required on each bank, and precautions should be taken to prevent soil erosion around the posts. Keep floodgates free of debris to prevent the water from backing up and flooding the adjacent land.

Fence Maintenance

For longer and better service from your fences, keep them in good condition. Spring and fall are the best times to inspect and repair fences.

Reset, repair, or replace anchor posts (gate, corner, end, and braced-line posts) whenever needed. Don't neglect line posts, but they are not quite so important. Refasten loose wires to the posts, and splice broken wires back together to prevent further damage.

Keep the fencing or wires properly stretched. You can take up the slack in individual wires by adding tension curves, using the crimper tool shown earlier.

Keep the fence line cleared of weeds and brush. Such growth not only detracts from the appearance of the fence, but it can be an added fire hazard.

Climb your fences only when absolutely necessary. Repeated climbing, particularly at the same point, may weaken or damage the fence. Install a gate or stile at the point where frequent crossing is necessary.

Lightning Protection

Livestock may be killed instantly if near an ungrounded or improperly grounded wire fence that receives a lightning discharge. The current may travel as far as 2 miles along the wire fence.

Wire fences attached to trees and buildings are most likely

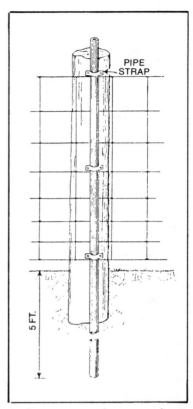

Grounding of Woven Wire With Fence Posts

to receive lightning discharges, but any wire fence can be a hazard.

Proper grounding of a fence greatly reduces the chance of electrocution, but, because lightning is so highly erratic, there will always be some danger.

All-steel fenceposts will ground a wire fence if they are in contact with wet or moist soil. But unless you can be sure that your posts are in contact with wet soil, you should either use extra long posts set at least 5 ft. in the ground or install special ground rods as recommended for wood posts below

Shown is the recommended method of grounding a wire fence with wood posts. Drive the ½- or ¾-in. steel rod at least 5 ft. into the ground and allow it to extend a few inches above the post. Fasten the rod to the post with the pipe straps so that it touches all the wires. Ground the fence in this manner every 150 ft.

Another way to ground the fence is to use an extra long steel post, set 5 ft. into the ground, every 150 ft.

Fence Building Machines

Fence-building machines are available for large-scale fencing, such as may be done on large farms or ranches. Some owner-operators of these machines may do custom fencing.

These machines are designed to build woven wire, barbed wire, and combination woven wire and barbed wire fences. Operations performed include driving the post, and unrolling, stretching, splicing, and attaching the fencing or wire.

With one of these machines, a three-man crew can erect about 1 mile of barbed wire fence or ½ mile of combination woven wire and barbed wire fence in a working day.

19. Caulks, Sealants, Putties and Glazing Compounds

Sealing your building against moisture and cold is a specialized business today boasting a variety of sophisticated sealants, caulks and glazing compounds, as well as conventional oil putties. The proliferation of sealing products calls for a little extra care in selecting a sealant, but the careful builder will be rewarded by finding just the right product.

Selection

Two of the best sources of information on the properties and application techniques of available sealing products are (1) your reputable paint, building supply or hardware store dealer, and (2) the labels on the products themselves. Here is a quick survey of some of the basic types of sealing compounds, with notes on advantages and disadvantages of each.

Putty

The name putty is usually applied to a soft, doughlike, knife-applied compound created by the blending of pigments and oils. It is used for face glazing (the application of a putty around the edges of a pane of glass), for bedding (the application of a compound to cushion the glass, applied to the inner side of the sash), and for filling crevices and cracks. Putties come in different types—for wood or metal and for interior and exterior use. Putties generally oxidize, aiding adhesion to the surface, and eventually hardening the compound. Once the putty has set, painting will extend its life.

Glazing Compounds

Glazing compounds, although similar to putties, differ in a special way. They are modified to enable them to remain plastic and resilient over a longer period of time than the conventional hard-setting putties. They are highly effective as a seal between glass and framing. Many different types, formulated for specific purposes, are available. A general-purpose glazing compound will bed and face glaze glass to wood and metal frames.

Oil-base Caulks

Oil-base caulks are designed to seal nonmoving cracks and joints in various materials of construction. They are soft and plastic, and consist of oxidizing oils, pigments and additives. They are low cost, form a surface skin overnight, and become firm usually in less than one year. Oil-base caulks can be used in filling joints where little or no expansion

and contraction is apt to occur. Life expectancy in joints with reduced activity is three to eight years if properly applied. Oil-base caulks are most commonly sold in a cartridge form, although they are also available in cans and pails, as well as in collapsible tubes. They are usually applied with a caulking gun. Ultimate shrinkage is usually five to twenty percent.

Coverage of any caulk or sealant is based upon the volume it will fill. Consequently, the heavier cartridge is usually not the better buy.

The life of an oil-base caulk can be prolonged with paint. Generally, oil-base caulks should be allowed to skin or cure before being painted. Because of the variety of types on the market, it's always best to consult the manufacturer's label for painting instructions.

Flexible Sealants

Flexible sealants are noted for their "stretchability." They are characterized by a good facility for elongation and a limited facility for recovery over a long period of time. These qualities make them practical for use in joints where hard-drying caulks would not be feasible. The most common of the flexible sealants are the butyls.

Butyl Sealants. Butyl sealants have a life expectancy of ten to fifteen years and are generally nonstaining to most materials. Gun grade butyl sealants become tack free in two to seventy-two hours and can then be painted over; they set firmly in about two weeks. They can be used in glazing and bedding applications and in joints where moderate movement may occur. This sealant stretches well but has a limited ability to recover from elongation. Use should therefore be limited to bedding of glass and filling of openings, both sides of which are of the same construction material. Butyl sealants in caulking form will normally shrink from 10 percent to 35 percent. depending upon the product, and thus are not recommended in joints larger than ¼ in. wide or deep.

Water-base Sealants

Acrylic Latex. Acrylic latex caulks or sealants are easy to apply and great timesavers. They can be applied to damp surfaces, tolled with water and painted almost immediately after application. Curing occurs quickly by water evaporation, leaving a sealant that is flexible and highly resistant. Water-base acrylic sealants adhere well to most surfaces. They are available in white and colors for use inside or out. Acrylics are recommended for all types of building joints but should not be used on traffic surfaces or where the joints would be continually submerged in water.

Polyvinyl Acetate Latex. Polyvinyl acetate latex-base caulks are generally recommended for interior use. These caulks dry very hard and are quite brittle. When used outdoors, these latex caulks can have problems with rewettability or washout if wet or rained upon before fully set.

Among the latex caulks, the acrylic latex sealants normally cost more than the polyvinyl acetates.

Solvent-base Acrylics

Solvent-base acrylics adhere well to almost any surface, and do not require primers. Life expectancy is in the 20-year range. Various colors are available, with good color retention.

Solvent-base acrylics are one-part, and develop a full cure in three to four weeks. The product is usually sold in cartridge form and heated prior to application.

Solvent-base acrylics have a strong odor, and for that reason should not be used in enclosed areas that are occupied, nor where odors might come in contact with food products. Solvent-base acrylic sealants, like the acrylic latex types, are recommended for all types of building joints, but should not be used on traffic surfaces or where the joints would be continually submerged in water.

Elastomeric Sealants

Elastomeric sealants have a rubberlike consistency, with a high degree of resiliency, weathering ability and tenacity. Elastomeric sealants have a longer life expectancy and are better able to adjust to movement in modern construction methods than are oil-base caulks. Therefore, these sealants are becoming increasingly popular with do-it-yourselfers.

Elastomerics are available in two forms: "one-part," which is used as supplied; and "two-part," which requires the mixing of a base compound with an "accelerator" immediately before use. Two-part sealants are seldom used by the homeowner. They are more complicated to work with than one-part elastomerics and are designed primarily to withstand the architectural stresses that occur in commercial buildings. Both types are usually supplied in gun grade consistency. The one-part almost always comes in caulking cartridges and the two-part in cans or pails. The following elastomerics are among the most widely used by homeowners today:

Hypalon. Hypalon sealants have excellent color stability and are available in a wide range of colors. They become tack free in one to two days and develop full cure in 60 to 90 days depending on temperature and width and depth of application. Hypalon can be used around windows and also in control and expansion joints. It requires no primer or special surface preparation, and cures to a rubber seal having excellent elongation and recovery. Hypalon has a life expectancy of 15 to 20 years.

Neoprene. Neoprene and nitrile one-part sealants have excellent elastomeric properties combined with good oil, chemical and heat resistance. Tack free time is one hour and curing time is about thirty days. These caulks do have high shrinkage and only fair color stability. Neoprenes are especially good for sealing driveway cracks and wall cracks before painting. Nitriles are recommended for small cracks and joints in metal frames and gutters. Be careful of the strong odor of nitrile when working in confined areas. Life expectancy is 15 to 20 years.

Polysulfides. Polysulfides have excellent durability and good adhesion to most surfaces. These properties can be enhanced by the use of a primer. The better grades have a life expectancy in excess of 20 years. Polysulfides are available in one-part and two-part systems. One-part systems are tack free in 24 to 48 hours, developing full cure in 30 to 90 days, depending upon temperature, humidity, and the width and depth of the application. Two-part systems are tack free in about 24 hours and develop full cure in 4 to 7 days, depending on temperature and humidity. One-part systems are not normally recommended for joints greater than ¾ in. wide and ⅜ in. deep.

Silicones

Silicones are one-part sealants unaffected by wide temperature ranges and have a life expectancy of about 20 years. They become tack free in less than 24 hours and cure in 2 to 5 days. Their chemical curing action can affect some substrates. Using a primer may bring about improved adhesion on some substrates, particularly on porous surfaces.

Caution should be used when choosing a silicone to be used near surfaces that are to be painted. Most silicones affect paint adhesion and cannot be painted.

SECTION IV: Maintenance

Where To Caulk

There are many areas in a building that may require caulking at one time or another. Here's a checklist that can help you keep up with your farm structure's caulking needs:

- Wall joints and cracks
- Joints between dissimilar materials
- Wall-to-slab joints
- Loose siding joints
- Stucco cracks
- Joints in concrete steps
- Around sinks and tubs
- Ceramic-tile joints
- Plumbing joints and fixtures
- Pipes through walls and floors
- Window frames
- Window glazing
- Flashing
- Sealing skylights
- Sealing ducts and vents
- Around air conditioners
- Leaky gutters and downspouts
- Door frames
- Column bases
- Around electric boxes

Preparation and Application of Sealants

Preparation for Caulking

To achieve a really effective and professional-looking caulking job, adequate preparation is a must. This is particularly true if you are planning to use an elastomeric sealant. Because of the contraction and expansion of the elastomeric sealant, good adhesion to the surface is essential; it cannot be obtained without surface preparation.

Careful clean-up is the first step. Moist, dusty or greasy surfaces must be cleaned or sealants will not adhere satisfactorily. To remove grease, just use a rag soaked in mineral spirits.

It is also necessary to allow enough space for a healthy amount of the compound. All joints should be at least ¼ in. wide and deep. If the area to be caulked is not large enough it must be enlarged before caulking. Too large a bead can also be harmful. Consult the label.

Removal of old putty and caulk that has become dried out or chipped is important. Scrape out all the old caulking with a putty knife and dust with a stiff bristle brush.

All wood surfaces should be primed. Steel sashes should be primed with a metal, rust-inhibitive priming material. Aluminum, or other nonrusting metals, should have all grease and other foreign matter removed.

Application of Caulking Compounds or Sealants

Gun Grade. This type of sealant is usually packaged in a cartridge which fits into a caulking gun. The compound is forced out in a uniform bead of caulking when the gun is triggered. Gun grade sealants are also available in cans or collapsible tubes. Caulk in collapsible tube form is useful in areas which are inaccessible to a gun or for small jobs. Tubes should be rolled up as the caulk is used to prevent skinning and to aid applying a steady bead.

When caulking with a cartridge, the nozzle of the cartridge should be cut at an angle and in a size appropriate to the opening to be filled. Puncture the inner seal. When applying the sealant, hold the point of the cartridge at an angle of approximately 45 degrees to the surface. Be sure to hold the gun parallel to the crevice—not at a right angle, and pull the gun along the crevice. Disengage the rod of the gun and pull it back to stop the flow of the caulk. Seal the nozzle of the cartridge with masking tape prior to storage.

Rope Form. Rope form caulking compounds are among the easiest for the do-it-yourselfer to apply. They are available in various dimensions and come coiled in rolls. They are suitable as temporary seals. Rope form caulks are particularly useful in providing protection against drafts, moisture or dust around windows, screens, louvres and miscellaneous points around a structure, particularly in a home.

No tools are required. Just pull one or more strands apart with your fingers and push them into the crevice to be filled. Because these rope form caulks are nondrying, they do not readily accept paint.

Use of Caulks, Sealants and Other Compounds

Caulking Tips

1. Buy all materials from a reputable dealer and follow the manufacturer's instructions closely as to the selection of the compound itself, its uses, and the method of application.

2. Caulking compounds should be used as prepared by the manufacturer. They should not be thinned with oil or thinners, because shrinkage and/or loss of adhesion may occur.

3. Check the manufacturer's instructions carefully to see if a primer is recommended.

4. In deep joints, back-up filler should be used; check to see what the manufacturer has recommended. Generally, if the joint is over a ½ in. deep, a filler is recommended prior to caulking. Oakum is not recommended as a filler; expanded polyethylene, polyurethane and nonstaining rubber fillers are recommended.

5. Most caulking compounds can be painted. Oil-base caulks should be painted whenever possible, as painting will extend their life expectancy. Always allow adequate setting time before painting.

6. When using a putty knife, remember to apply adequate pressure to get a smooth bead.

7. Avoid working at temperatures below 40° F. Below this temperature condensed moisture or frost may be present and prevent good bonding. Lower temperatures also tend to make the plastic materials less pliable and harder to handle.

8. For some jobs, such as caulking between dissimilar materials of construction, one of the newer and more expensive elastomeric sealants are recommended. On other jobs—those that are temporary or in places where joint movement is small—one of the less expensive caulks may be adequate.

Preparation for Glazing

Because of the complexities of sash design and the different characteristics of many putties and glazing compounds available, it is always wise to check the manufacturer's label thoroughly before you begin glazing. Check, too, to make sure that all components of the putty or glazing compound are blended thoroughly. Sometimes pigment will settle to the bottom of the container or liquid will rise to the surface; in these cases the can should be put on a paint shaker at the paint store.

Before face glazing a wooden sash, prime it with a good quality exterior paint primer or paint. When the paint has dried, apply a quantity of the putty or glazing compound to the sightlines of the glazing rabbet or groove.

Steps in Proper Glazing

Press the glass gently into the rabbet, leaving a bed of back putty or compound approximately ⅛ in. deep. Next apply glazing points or clips to hold the glass in place, and strip surplus putty or glazing compound at an angle to allow for proper run off of moisture condensation. Apply additional glazing compound or putty to the face and tool into place with a putty knife. Be sure to apply enough pressure here to fill the void. Using a putty knife again, trim the excess putty off. Allow the putty or glazing compound to remain untouched until it has attained a surface skin or become relatively firm. Then you may paint the putty or glazing compound and sash with one or two coats of paint. Paint will extend the life of the compound.

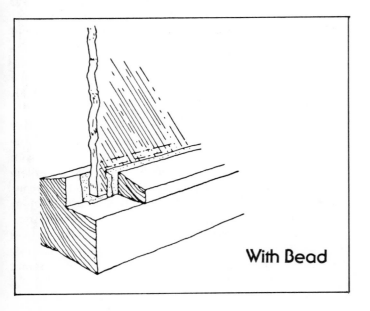

With Bead

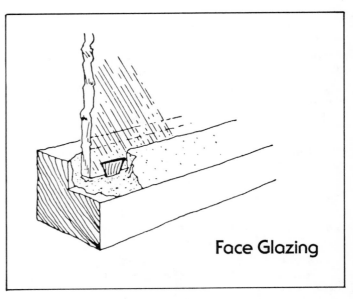

Face Glazing

20. Building Maintenance & Exterior Paint

Repairs

The cost to protect a building from serious decay can be minimal if buildings are inspected regularly and trouble is corrected early. At least once a year, areas and items most vulnerable to wetting should be inspected. Particular attention should be paid to roofs, roof edges (facia, soffits, rafter ends), joints in and adjacent to window and door frames, and appendages such as porches, steps, and rails. Any signs of repeated wetting and traces of decay should be investigated.

The crawl space, though it can be difficult to inspect, should not be neglected. In milder climates it is vulnerable to decay by the water-conducting type whose presence may or may not be revealed by strandlike surface growths. In colder climates it is subject to wetting by winter condensation, especially in the corners. During the summer in warm humid climates the floor may become wet from condensation created by summer air conditioning. The framing in the crawl space merits particular attention because it supports the weight of the building. Evidence of plumbing leaks should be searched for and any leaks promptly corrected.

Inspection for winter condensation in the crawl space should be made in the late fall and winter, and for condensation from air conditioning in the late summer. Check for wetting caused by air conditioning, particularly on the subfloor. Cupping or buckling of the finish floor is a sign that condensation may be occurring.

Inside the building, places to watch most carefully are shower rooms for water leakage into walls or floor, sink areas for plumbing leaks, and cold-storage for condensation within the walls, floor, or ceiling. Finding the source of a leak is sometimes baffling, because water may travel some distance from the point of leakage within the walls or floors near outside walls and damp-appearing surfaces indicate leaking water. Often a damp surface will be accompanied by molding. Steam radiators should also be routinely checked for leaks.

A moisture meter is helpful in locating wet zones within walls or in similar hidden places. A type of meter employing nonpenetrating electrodes can be used to measure moisture content of plaster and finish surfaces.

Recognition of Decay and Serious Wetting

Persistently wet places should be noted, and the cause corrected promptly. Various indications of wetting and decay are shown. Advanced decay is easily recognized; early decay may not be. Yet even a small amount of decay is cause for concern because serious strength losses usually accompany it. Thus, those who build or maintain wood structures should be acquainted with the ordinary signs of decay infection, particularly in load-bearing parts.

Discoloration of the wood. As decay progresses it usually imparts an abnormal color to wood. This change in color can be a useful diagnostic sign of decay if the inspector is reasonably familiar with the color or color shades of the sound wood. On surfaced wood the discoloration commonly shows as some shade of brown deeper than that of the sound wood. Some decays, however, produce a lighter than normal shade of brown, and this change may progress to a point where the surface might be called white or bleached. If this bleaching is accompanied by fine black lines, "zone lines," decay is virtually certain. Often, an abnormal variation in color creating a mottled appearance is more helpful in detecting early decay than actual hue or shade of discoloration. Highly indicative of decay, and especially conspicuous, is variable bleaching on a dark background of blue stain or mold.

Accompanying the color change, there may be an absence of normal sheen on the surface of infected wood. Here also, familiarity with the normal appearance of the wood can be of great help in recognizing the loss of sheen. Occasionally, in relatively damp situations, the presence of decay infection will be denoted by surface growth of the attacking fungus; in these cases the wood beneath usually is weakened, at least superficially.

Stain showing through paint films, particularly on exterior woodwork, is evidence of serious wetting and probable decay beneath the film. Rust around nail heads suggests that wetting has been sufficient for decay to occur.

Loss of wood toughness and hardness. Wood can also be examined for decay by simple tests for toughness of the fibers and for hardness. Toughness is the strength property most severely reduced by early decay. The pick test is a helpful and widely used simple means of detecting diminished toughness. It is made most reliably on wet wood. An ice pick, small chisel, sharpened screwdriver, or similar sharp-pointed or edged instrument of tough steel is jabbed a short distance into the wood; and a sliver pried out of the surface. The resistance offered by the wet wood to prying and the character of the sliver when if finally breaks are indicative of toughness.

In the pick test, sound wood tends to break out as one or two relatively long slivers and the breaks are of a splintering type. Where loss of toughness has been appreciable, the wood tends to lift out with less than usual resistance and usually as two relatively short pieces. Moreover, these short pieces break at points of fracture; that is, abruptly across the grain with virtually no small splinters protruding into the fracture zone.

On planed lumber, reduced toughness of wood from early decay is sometimes indicated by abnormally rough or fibrous surfaces. Similarly, the end grain of a board or timber may be rougher than usual after sawing.

Toughness may also be reduced by certain other factors such as wood compression, wood tension, or compression failures. There usually is little doubt of decay infection if the weakening is accompanied by a decay-induced discoloration.

In many cases the reduced hardness of infected wood can be detected by prodding the wood with a sharp tool. Softening, however, usually is not so obvious nor so easily detected at early stages as decreased toughness.

Shrinkage and collapse. Decay in the more advanced stages frequently causes wood to shrink and collapse. Under paint, this may be first manifested by a depression in the surface. Often the paint will acquire a brown-to-black discoloration or loosening of paint, particularly at joints.

Surface growths. Decay in crawl spaces invaded by a water-conducting fungus may be evidenced by fanlike growths, vinelike strands, or by a sunken wood surface resting on foundation walls or piers. Such decay is usually most advanced near the foundation, because the fungus usually starts there. The fanlike growths are papery, of a dirty white with a yellow tinge. They may spread over the surface of moist wood or —more commonly—between sub- and finish-flooring or between joists and subfloor. These growths may further appear under carpets, in cupboards, or in other protected places. Water-conducting, vinelike strands grow over the foundation, framing, the underside of flooring, inside hollow concrete blocks, or in wall voids.

The fungus carries water through these strands from the damp ground or other source to the normally dry wood being attacked. Usually the main water conductors are ¼ to ½ in. wide although they sometimes reach 2 in. They are similar in color to the fanlike growths although they can turn brown to black. During dry weather, shrinkage cracks in floors often outline the extent of an attack. Rotted joists and subflooring in relatively dry crawl spaces usually have a sound appearance even when the interior wood is essentially destroyed.

Corrective Measures

The cause of excessive wetting of any wood members in a building should be investigated, and measures to correct the situation should be taken as soon as practicable. Ordinarily, alterations or repairs to stop the wetting and to keep the damaged item dry are sufficient. If there is any question about whether or not the moisture source has been eliminated, the replacement wood should be treated with preservative. For preventing decay-producing situations, it is important to remember that wood will not decay if its moisture content is no more than about 20 per cent. To decay it must be actually in contact with water; moisture imparted by damp air alone can cause objectionable swelling, but cannot support decay.

In making repairs required because of decay, it usually is necessary to replace only wood so weakened that it is no longer serviceable. Infected wood will not endanger adjacently placed sound wood so long as both are kept dry.

Because of the high decay hazard of the roof edge, it should be watched for signs of wetting and decay. With flat roofs, gravel stops cannot be kept watertight for appreciable lengths of time unless regular attention is given to the joint seals. These should be resealed at the first sign of leakage. If eave gutters show corrosion, they should be covered with a corrosion-inhibiting paint. This is particularly necessary with recessed gutters because considerable hidden decay can occur before leakage is evident. Where no edge flashing has been used, it should be installed if there is evidence of sufficient wetting to cause staining of the last sheathing board, rafter ends, or facia.

Maintenance by Preservative Treatment

New wood for replacing areas of decay in buildings should receive the same type of preservative treatment that would be recommended for new construction in those areas. Flooding treatment with brush or spray of items that have been in service for a considerable time is not likely to prevent or arrest decay if the job is so big that it must be done on a large scale. It can be effective, however, if it is a home operation where the owner can give the time needed to get the preservative solution sufficiently deep into joints and cracks.

In flood treatment the object is to get the preservative deep into joints and crevices where rainwater is likely to be trapped and to have the penetration equal that reached by the water. Repeated treatment every few years will add to the margin of safety. Examples of items that can be protected by flooring of items in place are: bases of porch pillars and carport posts resting on concrete, plank porch floors, shutters, window boxes, and lookouts. Where rain seepage into a joint is indicated by failure of nearby paint, trouble from this source can be minimized if the joint is flooded by spraying or brushing with a water-repellent preservative prior to repainting. When treating, the preservative should be kept off plants and grass and not permitted to accumulate on the skin. A suitable preservative for most maintenance is 5 per cent pentachlorophenol with water repellents dissolved in mineral spirits; it is widely available and is identified on the label.

Eradicating Water-Conducting Fungus

Decay caused by the water-conducting fungus is easily prevented by incorporating the building procedures that have been discussed in previous sections. However, if attack has already occurred, special control measures are required. If the fungus is well established and conditions supporting it are not removed, large areas of flooring or walls may have to be repeatedly replaced. Cases are reported in which replacements were necessary at 1- and 2-year intervals.

The water-conducting fungus is susceptible to drying; therefore, it can be permanently separated from its source of water. When this is accomplished, the affected wood soon dries, and the fungus dries within a few weeks. Then only wood that has been too weakened to safely support its load needs replacement. However, if there is any doubt that the fungus has been eliminated, it is safest to replace all infected wood with wood that has been pressure treated with a preservative.

Usually the fungus gets its water from the ground or, less frequently, from wet wood or masonry in the general area where the decay occurs. Following are the most common measures to control the water-conducting fungus; some will be recognized as measures that should have been taken at time of construction.

- Wherever the location of decay, seek out and remove any wood forms left from pouring concrete steps or foundations as well as any other wood, building paper, insulation board, or similar cellulosic material that may offer a direct bridge from the soil to the wood of the building. Also, eliminate stumps and all building debris. If necessary, regrade the crawl space or soil outside the building to provide wood-soil clearance.
- Provide for drainage of surface water away from the outside foundation and the crawl space. If the ground in the crawl space is not dusty dry, include better ventilation or a soil cover, with the aim of making the air in the space dry enough to restrain the fungus in its development of water-conducting strands. A polyethylene film sheet is better than roll roofing for a soil cover when the water-conducting fungus is present, because this fungus will attack asphaltic papers.
- Open the foundation of the porch and remove enough soil from the fill to expose the entire sill under the slab. The opening should be sufficient to permit inspection of the sill and provide ventilation to it. If the sill needs replacement, use pressure-treated wood. Termite-control operators are familiar with the technique of excavating fills.
- If water-conducting strands or other growths of the fungus are observed on concrete or brick foundations, scrape off the larger strands and brush off the remainder with a steel brush. Finally, the cleaned surfaces should be thoroughly flooded with a preservative. This treatment also is applicable to situations in which the fungus is found on concrete exposed by excavating fills or by removal of forms. Always examine the treated areas and re-treat if any evidence of new growth appears. Where it is apparent that water-conducting strands are hidden inside concrete blocks or in loose mortar in brickwork, insert a metal shield between the foundation and substructure wood or, if the construction is brick, reset a few upper courses using cement mortar.
- If the source of the decay is traced to a plumbing leak, repair the leak. Where the trouble is associated with shower stalls, a completely new watertight lining may be needed. If the framing and sheathing for the floor and walls of the shower are exposed during repairs, replace them with pressure-treated wood. The most dangerous leaks are the small ones that are difficult to detect.
- If attack occurs in a slab-on-ground supported building that does not meet waterproofing and ground-clearance standards, replace all basal plates with pressure-treated wood and use non-wood flooring. Provide as good outside clearance as possible and chemically treat slab edge and adjacent soil. Attack seldom occurs in slab-supported houses with adequate waterproofing of the slab and adequate ground clearance unless unusual wetting occurs, such as that caused by excessive lawn sprinkling or by elevated flower beds.
- If attack occurs in a basement, replace wood in contact with a wall or the floor with pressure-treated wood. Avoid having enclosed stairs, partitions finished on both sides, cupboards, or paneling on the outside walls in moist basements. This type of construction creates "dead air" spaces that promote growth of the fungus.
- If preservative-treated wood is specified, use creosote, pentachlorophenol, or a noncopper waterborne compound, applied under pressure. Although all heartwood of decay-resistant species is acceptable for some items and for exposures in well-designed new buildings, do not use such wood to replace wood that has been attacked by the water-conducting fungus. This fungus will attack the heartwood of most decay-resistant woods, including redwood and cedar.

Outdoor Painting

Regardless of the building materials used in your farm building, a painting make-over is an easy exercise.

During your inspection tour check some of the areas where painting problems are most likely to be brewing. These trouble spots include: window and door frames and surrounding areas; bases of columns on porches and entryways; steps; sidings; downspouts; under-eave areas; and in short, anywhere that moisture is likely to collect.

If excessive cracking or peeling of the old coating has occurred in these spots, or such conditions have spread over larger areas, removal of the paint will be necessary for a lasting paint job. If only spot-removing is called for, use a paint scraper to reduce the damaged surface to the bare wood, and then "feather" the edges of the area with a coarse-grained sandpaper.

If extensive moisture damage has occurred over a large surface area, however, it is best to obtain a sanding machine (available for rent in most cities). Using a scraper along with the electric sander you will find that this task can be easily completed. For best results, sand with the grain of the wood.

Exterior Paint and Other Finishes

Surface Types	Oil or Oil-Alkyd Paint	Cement Powder Paint	Exterior Clear Finish	Aluminum Paint	Wood Stain	Roof Coating	Trim Paint	Porch & Deck Paint	Primer or Undercoater	Metal Prime	Latex House Paint	Water Repellent
Wood Surfaces												
Clapboard	X.			X					X		X.	
Natural Wood Siding & Trim			X		X							
Shutters & Other Trim	X.						X.		X		X.	
Wood Frame Windows	X.			X			X.		X		X.	
Wood Porch Floor								X				
Wood Shingle Roof				X								X
Metal Surfaces												
Aluminum Windows	X.			X			X.			X	X.	
Steel Windows	X.			X.			X.			X	X.	
Metal Roof	X.									X	X.	
Metal Siding	X.			X.			X.			X	X.	
Copper Surfaces			X									
Galvanized Surfaces	X.			X.			X.			X	X.	
Iron Surfaces	X.			X.			X.			X	X.	
Miscellaneous												
Asbestos Cement	X.								X		X	
Brick	X.	X		X					X		X	X
Cement & Cinder Block	X.	X		X					X		X	
Concrete/Masonry Porches And Floors								X			X	
Coal Tar Felt Roof						X						
Stucco	X.	X.	X	X					X		X	

• dot at right of X indicates a primer or sealer may be needed before finishing coat is applied
SOURCE: U.S. Department of Commerce

Prepaint Preparation

If old coatings are slated for removal, another must is a prepainting prime job. Ask your paint dealer which primer is specifically formulated for the surface in question, and apply it like a top coat with a brush, a spraygun or a roller.

If during your inspection you find that scratches or cracks have appeared on some surfaces, these mars should be filled with compounds formulated for this purpose, and available at most paint stores and can be applied with a putty knife, kitchen utensil, or even your thumb. After filling the cracks, allow the compound to dry, and sand the area smooth before priming and painting.

If the condition of the surface is good, and the old coating is strong and clean, your pre-painting preparations will be minor. In such cases a thorough "going over" with a wire brush will remove the surface dirt. If, on the other hand, the exterior of your home is not as clean as it need be, a solution of household detergent or tri-sodium phosphate and water will do the job.

Acid cleansing products are available in a condensed form at most paint and hardware stores, but remember that all label instructions should be obeyed to the letter. Follow the scrub-down with an overall rinse with clear water.

The Fun Part of Painting

Assemble your tools, spread protective drop cloths, and you're ready for the fun part of painting.

Starting at the top and working your way down to ground level, apply a layer of exterior house paint in the color of your choice. Check with your dealer, and read labels to make sure your paint is best possible coating for the exterior building material.

A bit of caution is needed here, however; when using ladders to reach those high spots, keep safety in mind. Don't overreach! Move the ladder often, and you may save yourself from injury.

Now about brushes. If you are using an oil-based paint, a thinning agent will be required to clean up brushes or rollers after the coating job is done. A good pre-painting "break-in" can be accomplished by dipping the brush in linseed oil. But if a water-based paint will be used, clean up right under the tap, and no breaking in is required.

A 4½ to 5 in. brush is suggested for painting the larger areas of your home, while trim and the like should be coated with 1½ to 2 in. "sash" brush. Be sure to use only the tips of your brushes—not the sides—for applying the paint, and you will be able to use these tools again and again.

Smooth, even strokes should be used to work the paint into the surface, whether one or two coats will be applied.

The best time to paint the exterior of your home is during a relatively dry time of day when the temperature is above 40 degrees, and after morning dew has evaporated. If it is impossible to follow these instructions to the letter, however, some latex paints can be applied under slightly varying conditions. For instance, your surface may not need to be "bone dry" during application.

Choose Coatings Carefully

Choice of exterior paints should be geared to protection against weather, as well as to beauty and ease of application. Today water-based paints, often called latex, have been successfully geared to exterior use, which has cut down application and drying time considerably. But check with your dealer to assure yourself of the quality of any product you choose. Saving pennies on the original outlay for paint can mean wasting dollars in frequently needed repainting.

Wooden clapboard siding, one of the most commonly used exterior building materials, lends itself to almost any house paint formulated for wood surfaces.

Shingles of various decorative woods, on the other hand, may have a natural grain which is pleasing to the eye and can be coated with a clear water repellent preservative. Color can be added by applying two coats of a quality pigmented stain which will enhance beauty of the shingles, seal the surface, and provide protection against the weather.

Asphalt shingle siding requires a rather special treatment calling for exterior emulsions formulated for these surfaces.

Wooden trim, such as window sashes, shutters and doors, should be attractively coated with a colorful exterior enamel. These coatings, which dry with a relatively glossy surface, are available in either water- or oil-based mixtures, and in a variety of sheens. Those which have the smoothest surface are called "high-gloss" enamels, while others are classfied as "semi-gloss" coatings.

Masonry Surfaces

Masonry surfaces—brick, cement, stucco, cinder block, or asbestos-cement—can be revamped with a variety of paint products. One of the newest ideas in painting brick is a clear coating which withstands weather and yet allows the natural appearance of the surface to show through.

Cement-based paints are also used on masonry surfaces. Colorful rubber-based coatings, vinyl and alkyd emulsion paints, are also used on many types of masonry. Almost all exterior house paints may be applied to masonry, however, when surface preparations are made properly.

Metal Surfaces

Galvanized iron, tin or steel building materials are available in various types, all of which may rust if not protected against moisture. Copper building materials, although they will not rust, will give off a corrosive wash which will discolor surrounding areas. Aluminum, like copper, will not rust but will corrode if not protected.

Conventional house paints or exterior enamels can be applied to these surfaces. But there are some rust-inhibitive coatings which would be best suited for the job. Ask your paint dealer which paint is formulated for application to the metal used on the exterior of your building.

Porches, Floors, and Steps

Porch floors and steps are usually constructed of wood or concrete—which should be kept in mind when choosing paint. The most important point to remember, however, is that foot traffic on these areas is extremely heavy, so the paint must be durable.

Most paint stores stock special porch and deck paints which can wear well under this hard use, but the selection of a primer coat will vary according to the building material used. Wooden porches and steps, for instance, can be primed with a thinned version of the top coat, while cement areas may need to be primed wih an alkali-resistant primer.

Best results in painting concrete porches and steps can be obtained with a rubber-based coating, or similar product. Roughening the surface slightly with muriatic acid is recommended before painting concrete that is hard and glossy, but all label instructions should be closely followed.

Appendix A: Simple Methods of Wood Preservative Treatment
(Selected from U.S.D.A. Farmers' Bulletin No. 2049)

Treatments of Low First Cost

If you cut your own posts, have no help, and want to spend as little cash as possible, you may be willing to settle for a shorter post life—just so the treatment saves you money and work in the long run.

If so, preservatives such as water solutions of zinc chloride or chromated zinc chloride may work best for you, since the dry chemicals cost only 10 to 20 cents a pound and a pound will usually take care of one post. These chemicals are fairly clean to handle and are not dangerous to people or animals. They can be bought in dry form and mixed with water just before use, so that you get away from heavy freight charges. They can also be bought in the form of strong water solutions that call for extra water before using. These are easier to use and store than the dry form and may be practical where the freight haul from the supplier to you is not too long.

End-Diffusion Treatment on Green Unpeeled Posts

The end-diffusion or trough method of treating with zinc chloride is quite simple and has been given considerable study by the South Carolina Agricultural Experiment Station at Clemson College, the Forest Products Laboratory, and others. It consists in standing freshly cut unpeeled posts in a tub or other container into which you have poured a measured quantity of 15- to 20-percent zinc chloride solution or chromated zinc chloride solution. Copper sulfate is sometimes used but it is not recommended because it is highly corrosive to staples and fencing wire and has not done as well as zinc chloride in service.

About 5 pounds or approximately ½ gallon of 20-percent (by weight) zinc chloride solution is recommended for each cubic foot of post treated. The posts are allowed to stand with the butts down in the solution until approximately three-fourths of the solution has been absorbed—which may take from 1 to 10 days or longer. After treating with butts down the posts are turned over, and the tops are allowed to absorb the remaining solution. They are then stored for at least 30 days with the tops down to permit distribution of the preservative within the post before it is set in the soil.

The South Carolina Agricultural Experiment Station reports that the treatment works well with southern pine posts. The Forest Products Laboratory in Wisconsin has obtained fair to good penetrations and retentions in the treating of aspen and jack pine under the following conditions: (1) Posts cut during summer and early fall seasons, (2) treatment started within 7 days after cutting trees, (3) temperatures above freezing.

An average life of 8.5 years was estimated from 4 tests of birch and southern pine posts treated with either zinc chloride or copper sulfate by end diffusion while the average life of untreated posts of these woods was 4.4 years. Of 5 species treated with zinc chloride and tested in Mississippi, posts of slash pine and red oak were in serviceable condition after 4 years while those of sweetbay, sweetgum, and tupelo were estimated to have an average life of from 3 to 4 years.

The advantages and disadvantages of end-diffusion treatment can be summed up as follows:

Advantages
Low first cost.
Peeling not necessary.
Preservative may be bought and shipped dry or in concentrated form.
Requires little equipment.
Does not call for surplus of preservative.
Can be used conveniently for small batches of posts.

Disadvantages
Protection limited.
Results of treatment not uniform.
Water-borne preservatives subject to leaching.

Tire-Tube Treatment

Tank soaking treatments, except for end-diffusion treatments, are somewhat wasteful of preservative since they end up with leftover preservative. A treating method that gets you away from most of this loss is the so-called tire-tube treatment with water-dissolved preservatives as developed by the Forest Products Laboratory. This consists in setting posts on a slant, butt end up, with sections of old truck tire inner tubes slipped over the upper end, pouring a measured amount of zinc chloride solution into the supported tubes, and letting gravity force the solution lengthwise through the post, to replace the sap with treating solution. The method works only with green, round posts with the bark still on. It does not work with split or sawed posts. The posts preferably should be treated soon after cutting.[8]

The tire-tube treatment on posts of several species has provided an average life of 10 to 15 years.

Steeping Treatment

Another low-cost method of using water-dissolved zinc chloride to treat farm timbers is the simple tank steeping method. In this treatment either green or seasoned peeled posts or timbers are soaked for 1 or 2 weeks in unheated 5-percent zinc chloride solution. When time is very limited, the soaking period can be shortened to 3 days with fair results, but the longer treating times are better.

As with other methods of treatment, the results of steeping vary with different woods and with different exposure conditions. Posts of woods such as hickory, southern red oak, sweetbay, sweetgum, and water tupelo have not shown a significant increase in life as a result of steeping in zinc chloride, particularly when tested under warm moist climatic conditions. Southern yellow pine posts similarly treated perform somewhat better under these conditions. Western redcedar posts last reasonably long without treatment and no significant increase in service has been noted in posts of this species treated by steeping in zinc chloride. Treatment by steeping has been found to be definitely beneficial in the case of round posts of ash, jack pine, lodgepole pine, ponderosa pine, red pine, and Scotch pine tested in Wisconsin, Nebraska, and Montana. Posts of various species treated by steeping and included in 43 different installations have an estimated average life of 15.8 years.

Brushing

Brushing is also considered suitable only for preservative oils. It is generally advisable to apply two coats of the preservative by flooding rather than brushing it over the wood surface. The oil should be heated if it is not sufficiently fluid at the prevailing temperature. Special care should be taken to fill every check and crevice in the wood and to apply the preservative liberally to the end grain. The first coat should be allowed to dry completely before the second coat is put on.

The advantages of the brushing method over the others are its simplicity and the small quantity of preservative that is used. Sapwood pine 2 by 4 inches treated by brushing two coats (with creosote) showed a retention of 0.76 pound per cubic foot or about one-half that resulting from dipping 3 minutes in the same preservative. Brushing requires little equipment and you need have no excess of preservative left over after the wood is treated. Brushing is more time-consuming than dipping, however, and is therefore more costly. You will find it of value in treating parts of large sticks if they cannot be tank treated, and in treating timber at joints and all points of contact where decay is likely to occur. As mentioned before, brushing or dipping adds less to decay resistance than do the other treatments that result in real penetration.

Like dipping, the brushing method is suitable only for use on peeled, thoroughly seasoned, and dry timber. It is best to use it in warm weather. In cold weather, the preservative is cooled and penetrates less readily. Penetration in any event will be shallow.

The apparatus for treatment consists only of a pail and a suitable brush. A small soft broom might be substituted for the brush if desired.

Although brush applications of preservatives are often beneficial when the wood is not used where it is likely to be moist for long periods, service tests on posts indicate that the treatment has doubtful value when applied to wood used in contact with the ground. In 10 tests of posts of various species such as Douglas-fir, red oak, white oak, northern white-cedar, lodgepole pine, and spruce, the estimated average life of the treated posts is only 9.5 years while the average life of untreated posts of these woods is approximately the same. Of these 10 tests only 1 showed a definite increase in post life due to the brush treatment.

[a]Complete details of the tire-tube treatment are available in Report No. 1158, Tire-Tube Method of Fence Post Treatment, which can be obtained free from the Forest Products Laboratory, Madison, Wis. 53705.

Appendix A.1

Typical Footing Depths by Region for Federal Government-Approved Construction

REGION	DEPTH (in.)
Florida	6
Southern California	12
Tennessee	18
West Virginia	24
New Jersey	30
Ohio	32
Montana	36
Minnesota	42
Maine	48

Appendix B: List of Manufacturers and their Addresses

Alcan Aluminum Corporation
100 Erieview Plaza
Cleveland, Ohio 44114

Alsco Anaconda
Cascade Plaza I
Suite 2200
Akron, Ohio 44308

Aluminum Company of America
1501 Alcoa Building
Pittsburgh, Pennsylvania 15219

Amcord, Inc.
610 Newport Center Drive
Newport Beach, California 92660

American Plywood Association
1119 A
Tacoma, Washington 98401

American Standard, Inc.
40 West 40th Street
New York City, New York 10018

Andersen Corporation
Bayport, Minnesota 55003

Architectural Metals, Inc.
111 Pennsylvania Avenue
Paterson, New Jersey 07503

Armco Steel Corporation
703 Curtis
Middletown, Ohio 45043

Arrow Group Industries, Inc.
100 Alexander Avenue
Pompton Plains, New Jersey 91730

Atlantic Building Systems, Inc.
Post Office Box 1714
Atlanta, Georgia 30301

ATO, Inc.
4420 Sherwin Road
Willoughby, Ohio 44094

Bethlehem Steel Corporation
Bethlehem, Pennsylvania 18016

Bliss & Laughlin Industries
Oak Brook, Illinois 60521

Boise Cascade Corporation
One Jefferson Square
Boise, Idaho 83728

Brinkman Fence Company
Route 8
Huntington-Auburn Road
Topeka, Kansas 66604

Butler Brothers
Lewiston, Maine 04240

Butler Manufacturing Company
7400 East 13th Street
Kansas City, Missouri 64126

Cal-More Livestock Equipment, Inc.
28095 Front Street
Rancho, California 92390

Cascade Industries, Inc.
Talmadge Road
Edison, New Jersey 08817

The Ceco Corporation
5601 West 26th Street
Chicago, Illinois 60650

The Celotex Corporation
1500 North Dale Mabry C
Tampa, Florida 33607

Certain-Teed Products Corporation
Post Office Box 860
Valley Forge, Pennsylvania 19482

Champion International Corporation
777 Third Avenue
New York City, New York 10017

Coronis Framing Systems
960 Estate Street
Trenton, New Jersey 08604

Crown Aluminum Industries Corporation
Post Office Box 61
Roxboro, North Carolina 27573

Cyclops Corporation
650 Washington Road
Pittsburgh, Pennsylvania 15228

DAP, Inc.
Post Office Box 277
Dayton, Ohio 45401

Del-Tex Buildings
Box 19367
Houston, Texas 77024

Marshall Erdman and Associates, Inc.
5117 University Avenue
Madison, Wisconsin 53705

Evans Products Company
1121 S. W. Salmon Street
Portland, Oregon 97208

Flintkote Company
400 Westchester Avenue
White Plains, New York 10604

GAF Corporation
140 West 51st Street
New York, New York 10020

General Cable Corporation
500 West Putnam Avenue
Greenwich, Connecticut 06830

Georgia-Pacific Corporation
900 S. W. Fifth Avenue
Portland, Oregon 97204

W. R. Grace & Company
62 Whittemore Avenue
Cambridge, Massachusetts 02140

Hawkeye Steel Products, Inc.
Waterloo, Iowa 50704

Howmet Corporation
475 Steamboat Road
Greenwich, Connecticut 06830

Ideal Cement Company
821 17th Street
Denver, Colorado 80202

International Steel Company
1321 Edgar
Evansville, Indiana 47707

Johns-Manville Corporation
Greenwood Plaza
Denver, Colorado 80217

Jones & Laughlin Steel Corporation
3 Gateway Center
Pittsburgh, Pennsylvania 15230

Kaiser Aluminum & Chemical Corporation
300 Lakeside Drive
Oakland, California 94643

Kaiser Steel Corporation
Kaiser Center
300 Lakeside Drive
Oakland, California 94612

Keene Corporation
345 Park Avenue
New York, New York 10022

Keystone Constructions, Inc.
Peoria, Illinois 61607

King Pole Building Company
Post Office Box 35
Schenectady, New York 12304

Kinkead Industries, Inc.
5860 North Pulaski Road
Chicago, Illinois 60646

Kirby Building Systems, Inc.
Post Office Box 36429
Houston, Texas 77036

Koppers Company, Inc.
Forest Products Division
1900 Koppers Building
Pittsburgh, Pennsylvania 15219

Leach's Custom Corrals
19130 Walnut Drive
Rowland Heights, California 91748

Lear Siegler, Inc.
Cuckler Division
Post Office Box 346
Monticello, Iowa 52310

Lincoln Steel Corporation
Box 1668
Lincoln, Nebraska 68501

Macomber, Inc.
Post Office Box 8530
Canton, Ohio 44711

Masonite Corporation
29 North Wacker Drive
Chicago, Illinois 60606

Metal Building Manufacturers Association
1230 Keith Building
Cleveland, Ohio 44115

Morton Buildings
252 West Adams Street
Morton, Illinois 61550

National Association of Building Manufacturers
1619 Massachusetts Avenue NW
Washington, D.C. 20036

National Gypsum Company
325 Delaware Avenue
Buffalo, New York 14202

National Steel Corporation
Granite City Steel Division
20th and State
Granite City, Illinois 62040

Nelson Manufacturing Company, Inc.
Box 636
Cedar Rapids, Iowa 52406

OK Corrals & Equipment
25852 Springbrook Avenue
Saugus, California 91350

Overhead Door Corporation
6250 LBJ Freeway
Dallas, Texas 75240

Owens-Corning Fiberglas Corporation
Fiberglas Tower
Toledo, Ohio 43659

The Panel-Clip Company
Post Office Box 423
Farmington, Michigan 48024

Pascoe Steel Corporation
1301 East Lexington Avenue
Pomona, California 91766

Peoples Building and Supply Company
761 North Main Street
Hubbard, Ohio 44425

Pittsburgh Corning Corporation
800 Presque Isle Drive
Pittsburgh, Pennsylvania 15239

Pyrotronics, Inc.
8 Ridgedale Avenue
Cedar Knolls, New Jersey 07927

The Reasor Corporation
Post Office Box 460
Charleston, Illinois 61920

Reichhold Chemicals, Inc.
Reinforced Plastics Division
Post Office Box 81110
Cleveland, Ohio 44181

Republic Steel Corporation
Manufacturing Group
Albert Street
Youngstown, Ohio 44505

Rober IBG
Post Office Box 100
Wheeling, Illinois 60090

H. H. Robertson Company
Two Gateway Center
Pittsburgh, Pennsylvania 15222

Bill Ryan Enterprises
Post Office Box 205
Putnam, Connecticut 06260

Soule Steel Company
Post Office Box 3510
Rincon Annex
San Francisco, California 94119

Star Manufacturing Company
Post Office Box 94910
Oklahoma City, Oklahoma 73109

T. C. Ranch Company
2700 Pomona Boulevard
Pomona, California 91768

Umbaugh Pole Building Company
3777 West State
Route 37
Delaware, Ohio 43015

Umbaugh Pole Building Company
Post Office Box 300
State Route 146
Millbury, Massachusetts 01527

Umbaugh Pole Building Company
4833 Harding Avenue
Ravenna, Ohio 44266

United States Gypsum Company
101 South Wacker Drive
Chicago, Illinois 60606

United States Steel Corporation
600 Grant Street
Pittsburgh, Pennsylvania 15230

Jim Walter Corporation
1500 North Dale Mabry Highway
Tampa, Florida 33604

Warner Company
1721 Arch
Philadelphia, Pennsylvania 19103

Weyerhaeuser Company
Wood Products Group
Tacoma, Washington 98401

Wheeling-Pittsburgh Steel Corporation
4 Gateway Center
Pittsburgh, Pennsylvania

The Wickes Corporation
110 West A Street
San Diego, California 92101

Winnebago Industries
Forest City, Iowa 50436

Index

Adobe 98-101
Aluminum roofing 63-65
Animal housing 15-29
Annual costs 30
Arched roofs 44, 71, 76, 126
Asbestos cement shingles 58-59
Asphalt shingles 60-62
Automatic gate 146

Balloon frame 39
Barbed wire fences 168-169
Barns 22
   for horses 134-148
Battens 59, 89
Batter boards 32-34
Beams 39
   See also Rafters
Bedding material 22
Beef cattle 12, 16-17
   reaction to environment 52
Bin liner 78
Bituminous roofing 59
Board fences 169-171
Bolts 42
Box stalls 26
Bracing 42-43
   for frames 50
   for plywood 96
Brooder stoves, poultry 27
Building checklists 8-9
Builtup roofing 60

Cable 102-104
Cable fences 169
Cannibalism prevention 27
Canvas roofing 65
Cattle
   feeder 78
   fences 164
   guards 172
   manure 15
   pens 15, 38
   shelter 79
Cattle shute 149-150
Caulks 174, 176-177
Checklists 8-9
Concrete 35-38
Concrete lot system 18-20

Condensation 178
Conductors 102-104
Construction planning 30-31
Contour fencing 167
Controlled environments 51-54
   animal reaction to 51-52
Cords 103-104
Corrugated steel 70-81
   leaks in 77
   load table for 75
   projects with 78-81
   sheets 63-65
   siding 71, 74, 76
   storage 72-73

Dairy cows 12, 15, 51
Decay 178-180
Doors 43, 90-91, 100, 128
Downspouts 14, 66-68
Drainage tile 33
Ducks 28-29

Electrical boxes 104-106
Electrical plugs 106
Electrical receptacles/fixtures 109-110
Electric fences 171
Electricity 14, 102-112

Farmhouse 10-11
Farm shop 14, 132
Farrowing 21
   house 155, 157
Farrow to finish operations 18, 20-21
Feed
   for cattle 15-16
   for horses 22
   supply 16
   for swine 19
   weight of 12
Feed room 135, 137-138, 140
Fences 161-173
   construction of 164-172
   maintenance of 173
   types of 161
Fenceposts 162-167
   setting of 165, 167
Flashing 57-58, 60, 63, 65-66, 68
Flat roofs 44, 101

Flat-seam roof 62
Floodgate 173
Floors 34, 36-39
   plywood 91-93
   slotted 155-157
Fluorescent lamps 112
Food *See* Feed
Footing forms 34-35
Forage wagon 78
Foundations 34-37, 100
Frames 39-43
   for greenhouses 119
   pipe 150
   propagating 123-124
Fruits 12
Fungus 179-180
Fuses 108-109

Gable roof 22, 24, 44, 71
   determining area of 56
Gage 70
Galvanized steel 62-63
Gambrel roof 44, 71
   determining area of 56
Garage 12
Gates 145-148
   cattle and flood 172-173
Geese 28-29
Girts 39
Glazing 177
Glazing compounds 174
Grain 12
Greenhouses 113-131
   construction basics for 117-119
   design of 117
   environmental control in 120-122
   materials for 117, 119
   plans for 124-131
   types of 113-114
   watering in 121
Grounding 76-77, 173
Gutters 66-68

Handling alley 20
Hay 13
Heaters 120
Hip roofs 44
   determining area of 56

Hogs
   feeder for   153-155
   fences for   164
   housing site for   12
   shelter for   79-80
Horses   22-26
   fences for   164
Horse shelters   23-26, 134-148
   stalls   135, 137-138, 140-142
Hot beds   122-124
Humidity   32, 51-52

Incandescent lamps   111-112
Insulation
   control and materials   54
   for metal roofing   61
   with plywood   88, 97
   for poultry housing   27

Lamp shades/reflectors   110-111
Lapping   62-63, 74
Latches   148
Laying hens   52, 158-160
   floor system for   28
Layout of buildings   10-11
Lean-to structures   114-115
Lighting   110-112
Light sockets   110-111
Lumber   45-46, *See also* Wood

Machine storage   13
Manure   15, 19-20
Mason's cord   33-34
Metal roofing   61-65, 68
Monitor roofs   44

Nailing   50, 86
Nails   44-46, 88, 93, 96
   with corrugated steel   73
   basic information regarding   94
   in roofing   57, 68-69
   position of, for roofing   64
   See also Nailing

Offset Gable roofs   22
Open front buildings   22

Painting
   of adobe   100-101
   outdoor   180-182
   of plywood   94-95
   of steel   77-78
Pasture system   16, 18
Pens   20-21
Plate   39, 42
Platform frame   39-40

Plywood   82-97
   gussets   88
   indentification index   85
   joints   87, 90-91
   properties of   95-97
   types and grades of   83-85
Pole buildings   47-50
   for horses   139-140
   with plywood   87-88
Portable shute   78-79
Post frame   39, 41
Poultry   27-29
   fencing for   164
   housing for   12, 27, 132-133, 158-160
   portable shelter for   79-80
Preservative treatment   179, 183-184
Prevailing winds   10-12
Projects
   cattle shute   149-150
   chicken house   158-159
   greenhouse   124-131
   hog feeder   153-155
   horse barn   134-148
   movable shed   132-133
   portable one horse stable   136
   poultry brooder   160
   poultry feeder   158
   wagon rack   151-152
Propagating frame   123-124
Pump house   80
Purlins   39, 44, 73-75, 90
Putty   174

Rafters   44, 90
Repairs   178-180
Rigid frames   41
Roadside stand   78
Roll roofing   59-60
Roofing   55-69, 71-74, 101
Roofs   22, 55-69
   leaks in   68
   plywood   89-90
   types of   44-45
Roof slope   55, 61
Rust   70

Sanitation   27
Sealants   174-175
   application of   176-177
Shed roofs   22-23, 44, 71, 132-133
   determining area of   56
Sheep   12, 52
   fences for   164
Shelters   17, 23-26
Shingles   56-58
   asphalt   60-61
   lapping of   62-63
   strip   61
   repair of   68-69
Siding   71, 74-75, 86

Silage feeding   151
Sill   39, 41
Silo   12-13
Single wall construction   86-87
Sites for building   10-11, 32, 34
   greenhouse   113, 116-117
   horse barn   140
Slate shingles   58
Snow guards   66-68
Soil   32, 34-35
Solar heating   53-54
Space requirements for livestock   25-27
Split ring connectors   45
Standing seam   61-62
Stepped footings   33
Storage   13
Studs   39, 41-43
Subcontractors   30
Swine   18-19
   effects of feeding position on   153, 155
   fences for   164
   housing for   155-156
Switches   107-109

Tack room   22-23, 26, 135, 138
Temperature   32, 51-54
   control for greenhouses   122
Tie stalls   25
Tin roofing   61
Traffic lanes   22
Troughs   14, 101
Truck/wagon sideboards   78
Trusses   39, 45, 47-48, 88, 90

V-Crimp sheets   62-63, 65-66
Vapor barrier   54
Vegetable weights   12
Ventilation   51-53
   for greenhouses   120
   for horse barns   140, 144
   for poultry housing   27

Wagon rack   151-152
Water requirements   14
Weights of commodities   12, 97
Windbreaks   13, 80-81
Windows   43, 50, 100
Winds   32, *See also* Prevailing winds
Wire   102
Wiring   102-112
   exterior   119
Wood
   decay   178-180
   fenceposts   162-163
   floors   38
   shingles   56
   See also Lumber

Zinc coating   70

Other SUCCESSFUL Books

SUCCESSFUL PLANTERS, Orcutt. "Definitive book on container gardening." *Philadelphia Inquirer.* Build a planter, and use it for a room divider, a living wall, a kitchen herb garden, a centerpiece, a table, an aquarium—and don't settle for anything that looks homemade! Along with construction steps, there is advice on the best types of planters for individual plants, how to locate them for best sun and shade, and how to provide the best care to keep plants healthy and beautiful, inside or outside the home. 8½"x11"; 136 pp; over 200 photos and illustrations. Cloth $12.00. Paper $4.95.

BOOK OF SUCCESSFUL FIREPLACES, 20th ed., Lytle. The expanded, updated edition of the book that has been a standard of the trade for over 50 years—over a million copies sold! Advice is given on selecting from the many types of fireplaces available, on planning and adding fireplaces, on building fires, on constructing and using barbecues. Also includes new material on wood as a fuel, woodburning stoves, and energy savings. 8½"x11"; 128 pp; over 250 photos and illustrations. $5.95 Paper.

SUCCESSFUL ROOFING & SIDING, Reschke. "This well-illustrated and well-organized book offers many practical ideas for improving a home's exterior." *Library Journal.* Here is full information about dealing with contractors, plus instructions specific enough for the do-it-yourselfer. All topics, from carrying out a structural checkup to supplemental exterior work like dormers, insulation, and gutters, fully covered. Materials to suit all budgets and home styles are reviewed and evaluated. 8½"x11"; 160 pp; over 300 photos and illustrations. $5.95 Paper. (Main selection Popular Science and McGraw-Hill Book Clubs)

PRACTICAL & DECORATIVE CONCRETE, Wilde. "Spells it all out for you...is good for beginner or talented amateur..." *Detroit Sunday News.* Complete information for the layman on the use of concrete inside or outside the home. The author—Executive Director of the American Concrete Institute—gives instructions for the installation, maintenance, and repair of foundations, walkways, driveways, steps, embankments, fences, tree wells, patios, and also suggests "fun" projects. 8½"x11"; 144 pp; over 150 photos and illustrations. $12.00 Cloth. $4.95 Paper. (Featured alternate, Popular Science and McGraw-Hill Book Clubs)

SUCCESSFUL HOME ADDITIONS, Schram. For homeowners who want more room but would like to avoid the inconvenience and distress of moving, three types of home additions are discussed: garage conversion with carport added; bedroom, bathroom, sauna addition; major home renovation which includes the addition of a second-story master suite and family room. All these remodeling projects have been successfully completed and, from them, step-by-step coverage has been reported of almost all potential operations in adding on to a home. The straightforward presentation of information on materials, methods, and costs, as well as a glossary of terms, enables the homeowner to plan, arrange contracting, or take on some of the work personally in order to cut expenses. 8½"x11"; 144 pp; over 300 photos and illustrations. Cloth $12.00. Paper $5.95.

FINISHING OFF, Galvin. A book for both the new-home owner buying a "bonus space" house, and those who want to make use of previously unused areas of their homes. The author advises which jobs can be handled by the homeowner, and which should be contracted out. Projects include: putting in partitions and doors to create rooms; finishing off floors and walls and ceilings; converting attics and basements; designing kitchens and bathrooms, and installing fixtures and cabinets. Information is given for materials that best suit each job, with specifics on tools, costs, and building procedures. 8½"x11"; 144 pp; over 250 photos and illustrations. Cloth $12.00. Paper $5.95.

SUCCESSFUL FAMILY AND RECREATION ROOMS, Cornell. How to best use already finished rooms or convert spaces such as garage, basement, or attic into family/recreation rooms. Along with basics like lighting, ventilation, plumbing, and traffic patterns, the author discusses "mood setters" (color schemes, fireplaces, bars, etc.) and finishing details (flooring, wall covering, ceilings, built-ins, etc.) A special chapter gives quick ideas for problem areas. 8½"x11"; 144 pp; over 250 photos and diagrams. (Featured alternate for McGraw-Hill Book Clubs.) $12.00 Cloth. $4.95 Paper.

SUCCESSFUL HOME GREENHOUSES, Scheller. Instructions, complete with diagrams, for building all types of greenhouses. Among topics covered are: site location, climate control, drainage, ventilation, use of sun, auxiliary equipment, and maintenance. Charts provide characteristics and requirements of plants and greenhouse layouts are included in appendices. "One of the most completely detailed volumes of advice for those contemplating an investment in a greenhouse." *Publishers Weekly.* 8½"x11"; 136 pp; over 200 photos and diagrams. (Featured alternates of the Popular Science and McGraw-Hill Book Clubs). $12.00 Cloth. $4.95 Paper.

SUCCESSFUL SPACE SAVING AT HOME, Galvin. The conquest of inner space in apartments, whether tiny or ample, and homes, inside and out. Storage and built-in possibilities for all living areas, with a special section of illustrated tips from the professional space planners. 8½"x11"; 128 pp; over 150 B-W and color photographs and illustrations. $12.00 Cloth. $4.95 Paper.

SUCCESSFUL KITCHENS, 2nd ed., Galvin. Updated and revised edition of the book *Workbench* called "A thorough and thoroughly reliable guide to all phases of kitchen design and construction. Special features include how to draw up your own floor plan and cabinet arrangement, plus projects such as installing countertops, dishwashers, cabinets, flooring, lighting, and more. 8½"x11"; 144 pp; over 250 photos and illustrations, incl. color. Cloth $12.00. Paper $5.95.

SUCCESSFUL LIVING ROOMS, Hedden. A collection of projects to beautify and add enjoyment to living and dining areas. The homeowner will be able to build a bar, dramatize lighting, enhance or brighten up an old fireplace, build entertainment centers, and make structural changes. "The suggestions…are imaginative. A generous number of illustrations make the book easy to understand. Directions are concisely written…new ideas, superior presentation." *Library Journal.* 8½"x11"; 152 pp; over 200 illustrations and photos, incl. color. Cloth $12.00. Paper $5.95.

SUCCESSFUL LANDSCAPING, Felice. Tips and techniques on planning and caring for lawns, trees, shrubs, flower and vegetable gardens, and planting areas. "Profusely illustrated…this book can help those looking for advice on improving their home grounds. Thorough details." *Publishers Weekly.* "Comprehensive handbook." *American Institute of Landscape Architects.* Also covers building fences, decks, bird baths and feeders, plus climate-and-planting schedules, and a glossary of terms and chemical products. 8½"x11"; 128 pp; over 200 illustrations including color; $12.00 Cloth. $4.95 Paper.

IMPROVING THE OUTSIDE OF YOUR HOME, Schram. This complete guide to an attractive home exterior at low cost covers every element, from curb to chimney to rear fence. Emphasis is on house facade and attachments, with tips on enhancing natural settings and adding manmade features. Basic information on advantages or disadvantages of materials plus expert instructions make it easy to carry out repairs and improvements that increase the home's value and reduce its maintenance. 8½"x11"; 168 pp; over 250 illustrations including color; $12.00 Cloth. $5.95 Paper.

SUCCESSFUL LOG HOMES, Ritchie. Log homes are becoming increasingly popular—low cost, ease of construction and individuality being their main attractions. This manual tells how to work from scratch whether cutting or buying logs—or how to remodel an existing log structure—or how to build from a prepackaged kit. The author advises on best buys, site selection, evaluation of existing homes, and gives thorough instructions for building and repair. 8½"x11"; 168 pp; more than 200 illustrations including color. $12.00 Cloth. $5.95 Paper.

SUCCESSFUL SMALL FARMS—BUILDING PLANS & METHODS, Leavy. A comprehensive guide that enables the owner of a small farm to plan, construct, add to, or repair buildings at least expense and without disturbing his production. Emphasis is on projects the farmer can handle without a contractor, although advice is given on when and how to hire work out. Includes basics of farmstead layout, livestock housing, environmental controls, storage needs, fencing, building construction and preservation, and special needs. 8½"x11"; 192 pp; over 250 illustrations. $14.00 Cloth. $5.95 Paper.

SUCCESSFUL HOME REPAIR—WHEN *NOT* TO CALL THE CONTRACTOR. Anyone can cope with household repairs or emergencies using this detailed, clearly written book. The author offers tricks of the trade, recommendations on dealing with repair crises, and step-by-step repair instructions, as well as how to set up a preventive maintenance program. 8½"x11"; 144 pp; over 150 illustrations. $12.00 Cloth. $4.95 Paper.

OUTDOOR RECREATION PROJECTS, Bright. Transform you backyard into a relaxation or game area—without enormous expense—using the instructions in this book. There are small-scale projects such as putting greens, hot tubs, or children's play areas, plus more ambitious ventures including tennis courts and skating rinks. Regional differences are considered; recommendations on materials, construction methods are given as are estimated costs. "Will encourage you to build the patio you've always wanted, install a tennis court or boat dock, or construct playground equipment…Bright provides information on choosing tools, selecting lumber, and paving with concrete, brick or stone." *House Beautiful.* (Featured alternate Popular Science and McGraw-Hill Book Clubs). 8½"x11"; 160 pp; over 200 photos and illustrations including color. $12.00 Cloth. $5.95 Paper.

SUCCESSFUL WOOD BOOK—HOW TO CHOOSE, USE, AND FINISH EVERY TYPE OF WOOD, Bard. Here is the primer on wood—how to select it and use it effectively, efficiently, and safely—for all who want to panel a wall, build a house frame, make furniture, refinish a floor, or carry out any other project involving wood inside or outside the home. The author introduces the reader to wood varieties and their properties, describes major wood uses, advises on equipping a home shop, and covers techniques for working with wood including the use of paints and stains. 8½"x11"; 160 pp; over 250 illustrations including color. $12.00 Cloth. $5.95 Paper.

SUCCESSFUL PET HOMES, Mueller. "There are years worth of projects…The text is good and concise—all around, I am most impressed." *Roger Caras, Pets and Wildlife, CBS.* "A thoroughly delightful and helpful book for everyone who loves animals." *Syndicated reviewer, Lisa Oglesby.* Here is a new approach to keeping both pet owners and pets happy by choosing, buying, building functional but inexpensive houses, carriers, feeders, and play structures for dogs, cats, and birds. The concerned pet owner will find useful advice on providing for pet needs with the least wear and tear on the home. 8½"x11"; 116 pp; over 200 photos and illustrations. Cloth $12.00. $4.95 Paper.

HOW TO BUILD YOUR OWN HOME, Reschke. Construction methods and instructions for woodframe ranch, one-and-a-half story, two-story, and split level homes, with specific recommendations for materials and products. 8½"x11"; 336 pages; over 600 photographs, illustrations, and charts. (Main selection for McGraw-Hill's Engineers Book Club and Popular Science Book Club) $14.00 Cloth. $5.95 Paper.

BOOK OF SUCCESSFUL HOME PLANS. Published in cooperation with Home Planners, Inc.; designs by Richard B. Pollman. A collection of 226 outstanding home plans, plus information on standards and clearances as outlined in HUD's *Manual of Acceptable Practices*. 8½"x11"; 192 pp; over 500 illustrations. $12.00 Cloth. $4.95 Paper.

HOW TO CUT YOUR ENERGY BILLS, Derven and Nichols. A homeowner's guide designed not for just the fix-it person, but for everyone. Instructions on how to save money and fuel in all areas—lighting, appliances, insulation, caulking, and much more. If it's on your utility bill, you'll find it here. 8½"x11"; 136 pp; over 200 photographs and illustrations. $4.95 Paper.

WALL COVERINGS AND DECORATION, Banov. Describes and evaluates different types of papers, fabrics, foils and vinyls, and paneling. Chapters on art selection, principles of design and color. Complete installation instructions for all materials. 8½"x11"; 136 pp; over 150 B-W and color photographs and illustrations. $12.00 Cloth. $4.95 Paper.

BOOK OF SUCCESSFUL PAINTING, Banov. Everything about painting any surface, inside or outside. Includes surface preparation, paint selection and application, problems, and color in decorating. "Before dipping brush into paint, a few hours spent with this authoritative guide could head off disaster." *Publishers Weekly*. 8½"x11"; 114 pp; over 150 B-W and color photographs and illustrations. $12.00 Cloth. $4.95 Paper.

BOOK OF SUCCESSFUL BATHROOMS, Schram. Complete guide to remodeling or decorating a bathroom to suit individual needs and tastes. Materials are recommended that have more than one function, need no periodic refinishing, and fit into different budgets. Complete installation instructions. 8½"x11"; 128 pp; over 200 B-W and color photographs. (Chosen by Interior Design, Woman's How-to, and Popular Science Book Clubs) $12.00 Cloth. $4.95 Paper.

TOTAL HOME PROTECTION, Miller. How to make your home burglarproof, fireproof, accidentproof, termiteproof, windproof, and lightningproof. With specific instructions and product recommendations. 8½"x11"; 124 pp; over 150 photographs and illustrations. (Chosen by McGraw-Hill's Architects Book Club) $12.00 Cloth. $4.95 Paper.

BOOK OF SUCCESSFUL SWIMMING POOLS, Derven and Nichols. Everything the present or would-be pool owner should know, from what kind of pool he can afford and site location, to construction, energy savings, accessories and maintenance and safety. 8½"x11"; over 250 B-W and color photographs and illustrations; 128 pp. $12.00 Cloth. $4.95 Paper.

FINDING & FIXING THE OLDER HOME, Schram. Tells how to check for tell-tale signs of damage when looking for homes and how to appraise and finance them. Points out the particular problems found in older homes, with instructions on how to remedy them. 8½"x11"; 160 pp; over 200 photographs and illustrations. $4.95 Paper.